SpringerBriefs in Fire

Series Editor

James A. Milke

For further volumes:
http://www.springer.com/series/10476

Casey C. Grant

Interoperable Electronic Safety Equipment

Performance Requirements for Compatible and Interoperable Electronic Equipment for Emergency First Responders

Springer

Casey C. Grant
The Fire Protection Research Foundation
Quincy, MA
USA

ISSN 2193-6595   ISSN 2193-6609 (electronic)
ISBN 978-1-4614-8276-5 ISBN 978-1-4614-8277-2 (eBook)
DOI 10.1007/978-1-4614-8277-2
Springer New York Heidelberg Dordrecht London

Library of Congress Control Number: 2013942982

Printed on acid-free paper

Springer is part of Springer Science+Business Media (www.springer.com)

Preface

The fire service and other emergency first responders are currently benefiting from enhanced-existing and newly-developed electronic technologies for use with personal protective equipment (PPE) ensembles. In the past decade the rate of technological innovation has accelerated, and events such as those that occurred on 11 September 2001 have stimulated additional consideration of applications of this technology.

Examples of the application of this new technology are relatively common-place. One such example is the effort toward addressing chemical, biological, radiological, nuclear, and explosive (CBRNE) type events. Protective ensembles used by emergency first responders include or will soon include electronics such as communications, GPS and tracking, environmental sensing, physiological sensing, and other components now becoming practical solutions at emergency events.

How these technological components function in a single synergistic operating platform is of critical interest to fire fighter end users. For instance, they are already burdened by the sheer weight of all their personal protective gear, and carrying separate battery power supplies for each of their individual electronic components begs for reasonable logic.

Overall, the broad-scale integration and coordination of separate electronic-based equipment used by fire fighters in their personal protective ensembles is lacking. Today's fire fighters would directly benefit from a standardized platform/framework for their electronic safety equipment (ESE), and working toward this end point is important for the collective emergency response community.

The goal of this project is to develop performance requirements for the compatibility and interoperability of electronic equipment used by fire service and other emergency first responders. The project will achieve this goal through the following objectives:

- Develop an inventory of existing and emerging electronic equipment categorized by key areas of interest to the fire service.
- Document equipment performance requirements relevant to interoperability, including communications, power requirements, etc.
- Develop an action plan toward the development of requirements to meet the needs of emergency responders.

As a result of the information gathered throughout this project, the following recommendations have resulted:

1. **Moving Toward ESE Interoperability**

   1.1 **Supporting an Evolutionary Approach**. Promote concepts that support ESE platforms with individual components that are compatible, integrated, and interoperable. This would be an evolutionary path that recognizes the virtues of a centralized interoperable platform. An example is the combining and maximizing of the efficiency of various features (e.g., power supplies), which would potentially alleviate and mitigate other performance concerns (e.g., insufficient performance due to limited power supplies).

   1.2 **Related Professional Applications**. Identify and consider the lessons learned from professions with parallel ESE applications to structural fire fighting, such as aviation, military, space, and underwater diving.

2. **Establishing Central Concepts for ESE Interoperability**

   2.1 **Clarify Definition of ESE**. Clarify the definition of "ESE" to distinguish if it is intended to include or exclude portable, mobile, stationary, and/or field deployable equipment.

   2.2 **Define ESE Interoperability**. Define "ESE Interoperability" to distinguish it from fireground interoperability and wireless communication interoperability. A possible definition is: "ESE Interoperability—the ability of ESE to operate in synergy in the execution of assigned tasks."

   2.3 **ESE Categories**. Consider categorization of emergency responder ESE, such as:

   (a) Communications
   (b) Environmental monitoring
   (c) Physiological monitoring
   (d) Sensory support
   (e) Tracking/location

   2.4 **Responder Knowledge Base**. Continue to recognize, utilize, and support the Responder Knowledge Base as a mechanism for tracking available ESE.

   2.5 **Interoperability Performance Characteristics**. Consider the key interoperability performance characteristics for fire service ESE as electrical oriented and non-electrical oriented. Examples of electrical oriented performance characteristics include:

   (a) Inter-component communication
   (b) Centralized power supply and distribution
   (c) Non-interference
       Examples of non-electrical oriented performance characteristics include:

(a) Form, fit, and function
(b) Ergonomics
(c) User interface
(d) Donning and doffing

2.6 **Component Attributes**. Consider the primary ESE component attributes, which are:

(a) Operability
(b) Maintainability
(c) Durability
(d) Availability
(e) Stability
(f) Reliability

3. **ESE Interoperability Standardization**

3.1 **Standardize Interoperability Concepts**. Document interoperability concepts in consensus-developed codes and standards documents. Use these documented requirements and/or guidelines to provide an appropriate baseline to address the overall topic of interoperability.

3.2 **Define the Fire Service Landscape**. Better define the requirements for fire service ESE by clarifying fireground environments and fire fighter needs, with specific attention to how ESE will be used in different situations. Transpose this information into the requirements or guidelines in standardization documents.

3.3 **Consistency of Requirements**. Consistency of performance requirements across all emergency responder ESE is a sensible goal, and consideration of logical differences in performance requirements should be based on substantive technical rationale. Action items that should be considered include:

3.3.1 Revisit NFPA requirements for performance requirements for all ESE, using an approach similar to the recent analysis provided for PASS by the Intrinsic Safety Task Group for the NFPA ESE Technical Committee.

3.3.2 Consider this effort through the PPE Correlating Committee since it affects multiple Technical Committees under their direction.

3.4 **Periodic Re-Evaluation**. The performance characteristics for different ESE should be re-evaluated on a periodic basis, since the technological landscape is continually changing and subject to ongoing advancements that impact the respective requirements.

4. **Intrinsic Safety of ESE**

4.1 **Periodic Re-Evaluation**. The need for intrinsic safety requirements for different ESE should be re-evaluated on a periodic basis, since the technological landscape is continually changing and subject to on-going advancements that impact the respective requirements.

4.2 **Interoperability**. Consideration should be given to promote concepts of interoperability, since a centralized interoperable platform that combines and maximizes the efficiency of multiple power supplies would potentially also alleviate other concerns (e.g., the current trade-off for reduced power supplies to comply with more rigorous Division 1 requirements).

4.3 **Consistency of Requirements**. Consistency of intrinsic safety requirements across all emergency responder ESE is a sensible goal that should be founded on the inherent technological differences of ESE that justify different intrinsic safety requirements. Action items that should be considered include:

4.3.1 NFPA requirements for intrinsic safety should be revisited and considered for all ESE, using an approach similar to the recent analysis provided for PASS by the Intrinsic Safety Task Group for the NFPA ESE Technical Committee.

4.3.2 This effort should be considered by the PPE Correlating Committee since it affects multiple Technical Committees under their direction.

4.4 **Defining the Fire Service Landscape**. Better define the requirements for intrinsically safe ESE by clarifying fireground environments and fire fighter needs. Examples of factors that should be considered are:

4.4.1 Division 1 and Division 2 levels of safety are presently based on normal or abnormal probability of occurrence, but this is meant for the electrical equipment installed at a fixed location rather than portable equipment moving in and out of hazardous locations.

4.4.2 A likely scenario for fire fighters is a hazardous environment where it is not expected, such as a gas leak, and in those cases the equipment on-site (without intrinsic safety protection) would arguably introduce a more realistic danger than the fire fighters own equipment.

4.4.3 Since intrinsic safety is generally applicable across all applications, clarify the unique features of a fireground with other applications (e.g., petrochemical).

4.5 **Ongoing Dialogue**. Further dialogue should be facilitated among the intrinsic safety product standards developers (i.e., FM, TIA, UL, etc.), manufacturers, and the user community, to clarify additional details needed for the proper references between the enabling standards (i.e., NFPA) and the product standards.

Project Technical Panel

Bob Athanas, FDNY/SAFE-IR (NY)

Christina Baxter, DOD/TSWG (VA)

Paul Greenberg, NASA Glenn Research Center (OH)

Bill Haskell, NIOSH NPPTL (MA)

Phil Mattson, DHS (DC)

Dan Rossos, Portland Fire & Rescue (OR)

Jeff Stull, International Personnel Protection (TX)

Dave Trebisacci, NFPA (MA)

Bruce Varner, Electronic Safety Equipment TC Chair (AZ)

Acknowledgments

Appreciation is expressed to all the participants who contributed their time to make this project a success. Special thanks are expressed to the National Fire Protection Association (through the NFPA Technical Committee on Electronic Safety Equipment) and Underwriters Laboratories for hosting the two project workshops. In particular, the Fire Protection Research Foundation expresses gratitude to the Fire Grants Program of the National Institute of Standards and Technology for providing the funding for the project and this workshop.

The content, opinions, and conclusions contained in this report are solely those of the author and do not necessarily represent the views of the Fire Protection Research Foundation. The Foundation makes no guaranty or warranty as to the accuracy or completeness of any information published herein.

Keywords: Electronic, Safety, Equipment, Electronic safety equipment, ESE, Emergency responders, Fire, Fire service, Interoperable, Interoperability, Compatible, Compatibility, Intrinsic safety, Communications, Radios

Contents

Chapter 1
Introduction and Background

The fire service and other emergency first responders are currently benefiting from enhanced-existing and newly-developed electronic technologies for use with personal protective equipment (PPE) ensembles. In the last decade the rate of technological innovation has accelerated, and events such as those that occurred on 11 September 2001 have stimulated additional consideration of applications of this technology.

Examples of the application of this new technology are relatively common-place. One such example is the effort toward addressing CBRNE (Chemical, Biological, Radiological, Nuclear, and Explosive) type events. Protective ensembles used by emergency first responders include or will soon include electronics such as communications, GPS and tracking, environmental sensing, physiological sensing, and other components now becoming practical solutions at emergency events.

How these technological components function in a single synergistic operating platform is of critical interest to fire fighter end users. For instance, they are already burdened by the sheer weight of all their personal protective gear, and carrying separate battery power supplies for each of their individual electronic components begs for reasonable logic.

Overall, the broadscale integration and coordination of separate electronic-based equipment used by fire fighters in their personal protective ensembles is lacking. Today's fire fighters would directly benefit from a standardized platform/framework for their electronic safety equipment (ESE), and working toward this end point is important for the collective emergency response community.

1.1 Need for Interoperability and Compatibility

Firefighting is a dangerous profession. In 2008, U.S. fire departments responded to an estimated 1,451,500 fires (based on the most recent year when full statistical data was available). These reported fires caused 3,320 civilian deaths, 16,705

C. C. Grant, *Interoperable Electronic Safety Equipment*,
SpringerBriefs in Fire, DOI: 10.1007/978-1-4614-8277-2_1,
© Fire Protection Research Foundation 2012

civilian injuries, $15.5 billion in direct property damage, 105 on-duty fire fighter fatalities, and 79,700 on-duty fire fighter injuries (Karter 2007; Fahy 2009; Karter 2008, 2009). In calculations of the total cost of fire, these losses translate into a combined total of $62 billion in 2008 (Hall 2011).

Based on 2007 data compiled in an NFPA profile report of the U.S. fire service, there are approximately 30,000 fire departments in the U.S. with roughly 1.1 million fire fighters. Just under three-quarters (73 %) of the 1.1 million fire fighters are volunteers, and nearly half of these volunteers serve in communities with a population of less than 2,500. Only one in 15 fire departments is all-career, but 43 % (or about two of every five) U.S. residents are protected by such a department. Approximately two-thirds of fire departments also handle emergency medical service (EMS) activities (Karter 2007).

For the emergency responders to remain effective, these electronic technologies must interact and operate synergistically and provide an effective and efficient overall package. Integration of these components with the emergency responder ensemble is required for managing weight, space, heat, and power requirements, as well as to create the least interference and burden to the equipment user. Figure 1.1 provides a symbolic depiction of the conceptual path of convergence for emergency responder ESE.

With the coming of age of the modern fire fighter's personal protective ensemble, it is imperative that this equipment work together for the ultimate benefit of the end user. Today's fire fighters are carrying an assortment of electronic equipment. Most of today's fire fighters are equipped with a flashlight, a personal alert safety system, and often a mobile radio. Other equipment not uncommon includes a thermal imaging camera, a hazardous gas analyzer, a video camera, or a heads-up display within the self-contained breathing apparatus (SCBA). What had once seemed like futuristic technological developments are

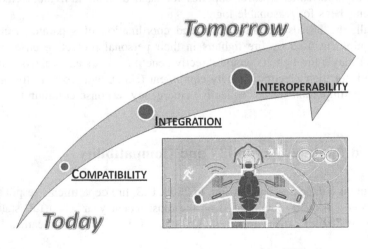

Fig. 1.1 Conceptual path of convergence for emergency responder ESE

instead rapidly emerging, such as position locating/tracking devices, head-mounted display augmented reality, environmental sensors, physiological monitors, and electronic textiles.

The operational characteristics of these electronic devices is typically not coordinated, and thus each has its own power supply, functional energy levels, input/output protocols, and so on. Devices are competing for space on the ensemble and power supply. Likewise they are competing with how they interface with the user, not only the direct wearer of the electronic equipment but also incident command, which may be receiving information from multiple fire fighters. This is especially important with the proliferation of wireless connectively, to ensure that equipment not only communicates properly back and forth with the user but also does not interfere with the operation of other equipment.

The electronic safety equipment serving today's fire fighters provides critical support to assist with the safe, efficient, and effective implementation of their duties. Yet a clear need exists to coordinate this equipment so that collectively it functions on a single operational platform.

1.2 Project Scope, Goal and Objectives

The topic of mobile electronic safety equipment used by emergency responders is far-reaching and continually evolving. Addressing the interoperability and compatibility of this equipment adds another layer of complexity. The scope of this research study is based on our understanding and definition of certain key aspects of this topic. Examples include what is and is not included with "mobile electronic safety equipment" and who are the "emergency responders" being addressed.

The goal of this project is to develop performance requirements for the compatibility and interoperability of electronic equipment used by fire service and other emergency first responders. The project will achieve this goal through the following objectives:

- Develop an inventory of existing and emerging electronic equipment categorized by key areas of interest to the fire service.
- Document equipment performance requirements relevant to interoperability, including communications, power requirements, etc.
- Develop an action plan toward the development of requirements to meet the needs of emergency responders.

In general, electronic safety equipment is understood to be all personal protective equipment (PPE) used by emergency responders that is electrically powered. The term "Electronic Safety Equipment" is defined as follows: "Products that contain electronics embedded in or associated with the product for use by emergency services personnel that provides enhanced safety functions for emergency services personnel and victims during emergency incident operations" (Sect. 3.3.6 Proposed Standard on Electronic Safety Equipment 2006).

Because this project is focused on the interoperability/compatibility of electronic safety equipment used by individual emergency responders, it should not be confused with concepts of "interoperability" and "compatibility" that are often used to discuss the strategic operation of large scale resources beyond the equipment used by individuals. For example, one study of this type reviews the interoperation of all emergency responder resources within the Washington DC area (Krimgold 2006). However, the study is organizationally focused and beyond the scope of this project, which is focused on the personal protective equipment used by an individual emergency responder.

This project is focused on the electronic safety equipment used by fire fighters in structural or proximity applications. Like other professions, the fire service has multiple roles and duties, and thus it is acknowledged that not all fire fighters are the same. For example, the diverse types of fire fighting duties represented by the various professional qualification standards available from NFPA are summarized in Fig. 1.2.

Further clarification of the different duties handled by modern fire fighters is illustrated in Fig. 1.3. This represents a layered composite of information representing various aspects of this professional community.

The most common model for organizations hosting U.S. fire departments is through local, regional/state, or federal based governments, though other

NFPA 1000	• Accreditation and Certification Systems
NFPA 1001	• Fire Fighter
NFPA 1002	• Fire Apparatus Driver/Operator
NFPA 1003	• Airport Fire Fighter
NFPA 1005	• Marine Fire Fighting for Land-Based Fire Fighters
NFPA 1006	• Technical Rescuer
NFPA 1021	• Fire Officer
NFPA 1026	• Incident Management Personnel
NFPA 1031	• Fire Inspector and Plan Examiner
NFPA 1033	• Fire Investigator
NFPA 1035	• Safety Educator, Info Officer, Firesetter Intervention Specialist
NFPA 1037	• Fire Marshal
NFPA 1041	• Fire Service Instructor
NFPA 1051	• Wildland Fire Fighter
NFPA 1061	• Public Safety Telecommunicator
NFPA 1071	• Emergency Vehicle Technician
NFPA 1081	• Industrial Fire Brigade Member
NFPA 1091	• Traffic Control Incident Management

Fig. 1.2 Diversity of fire fighter duties based on NFPA professional qualification standards

Fig. 1.3 Overview of the
U.S. fire service

arrangements exist, such as with private fire departments or in jurisdictions in Native American Nations (i.e., under tribal authority). The various duties and types of fire fighting are likewise indicated, as well as if the organization is based on career, volunteer, or combination staffing. In summary Fig. 1.3 represents an overview of the U.S. fire service and illustrates the organizations that administratively host fire departments, the basic types of fire service organizations, and types of operational functions that they typically handle.

Other electronic safety equipment used by fire fighters performing other types of fire fighting duties are considered beyond the scope of this study, which is primarily focused on structural fire fighting. For example, wildland fire fighters have unique requirements in that they can be deployed for weeks at a time in remote locations and thus have different power supply requirements for electronic gear. Another example are the portable, electronic tools used by fire marshals and fire investigators (e.g., computer tablets, cameras, etc.), which are required to handle different data in less hostile environments.

The tasks of the project include a review of the literature, an inventory of the equipment generally available to today's emergency responders, and a summation of the performance requirements for interoperability. In addition, this project includes documentation of a workshop to review and further refine this information. A summary of these project tasks are included in Fig. 1.4.

This research program is made possible through funding from the National institute of Standards and Technology (NIST) and has been conducted under the auspices of the Fire Protection Research Foundation. The project deliverables are intended for consideration by all who are interested in optimizing the electronic equipment for fire service end users. It's envisioned that this information will be

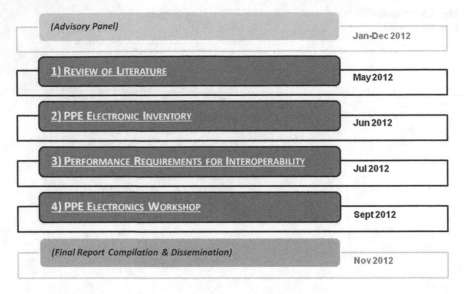

Fig. 1.4 Project tasks

particularly useful for manufacturers of fire fighter electronic safety equipment and
the relevant standards development committees of NFPA, ASTM, and others that
are continually trying to guide evolving technology in a manner to provide the
safest, most efficient, and most effective equipment for the end user.

Chapter 2
Literature Review

Supporting background information is provided on this topic through a review of the published literature. The literature addresses various characteristics of importance, including general interoperability concepts, applicable technological innovation, specific emergency responder equipment, model regulations, policy, and future trends.

2.1 Background and Methodology

The interoperability and compatibility of emergency responder electronic safety equipment is a relatively broad topic. Some sub-aspects of this overall topic have appreciable depth, such as the technological development of mobile electronic equipment with cell phones and other personal electronic equipment. On this basis the literature that has been included within this literature review is intended to support the further evolution of emergency responder electronic safety equipment and has been identified as the most appropriate literature for this purpose.

Based on an ongoing review of the applicable literature on this topic, several categories have evolved that provide a logical separation of this information. This categorization is summarized in Table 2.1, Categories of Literature Subtopics. This separation is intended to assist the user of this literature with further work as appropriate, and the subsequent review of the literature contained herein is conducted according to these categories.

The use of electronic safety equipment to perform job tasks is not unique to the fire service. Other professions have similar needs based on challenging work environments that require interoperability and compatibility of electronic safety equipment to optimize their mobile protection. Some of these professions have intensely focused on this topic in previous decades, and they possibly offer additional insight that may be of use to emergency responders.

The professional fields most similar to structural fire fighting include ground soldier military applications and Navy shipboard fire fighting, aviation fighter

C. C. Grant, *Interoperable Electronic Safety Equipment*,
SpringerBriefs in Fire, DOI: 10.1007/978-1-4614-8277-2_2,
© Fire Protection Research Foundation 2012

Table 2.1 Categories of literature subtopics

a. Background and methodology

b. Interoperability and compatibility of mobile electronics

c. General concepts for emergency responder equipment

d. Specific emergency responder equipment

e. Regulations, model documents, and policy

f. Future trends

Fig. 2.1 Examples of professions using interoperable ESE

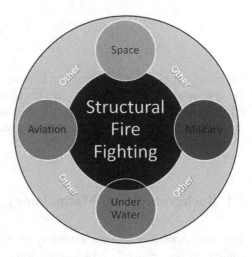

pilots, astronauts, and underwater diving applications. These are illustrated graphically in Fig. 2.1, Examples of Professions Using Interoperable ESE. During the conduct of this literature review, these other areas of professional focus are likewise considered.

Table 2.2, Comparison of Professions Using Interoperable ESE, provides a side-by-side comparison of the aforementioned applications that are using ESE. This illustrates the interoperability characteristics of the technology used in each particular arena, along with a subjective evaluation of the interoperability status, hazard exposure duration, and intensity of exposure.

2.2 Interoperability and Compatibility of Mobile Electronics

Mobile electronic equipment is an area of technology development that is experiencing intense focus. The marketplace for personal electronic equipment that's available to today's public is rapidly proliferating. In a competitive marketplace, the interoperability and compatibility of this equipment is of paramount

Table 2.2 Comparison of professions using interoperable ESE

	Application	Interoperability status	Hazard exposure duration	Intensity of exposure
Baseline	Structural fire fighting	Evolving	Short	Extreme
1	Aviation	Relatively mature	Moderate	Low
2	Military (e.g., ground forces)	Relatively mature	Short	Moderate
3	Space	Mature, well-proven	Long	Extremes
4	Underwater	Evolving	Moderate	Moderate

Fig. 2.2 Key electrical interoperability performance characteristics for ESE

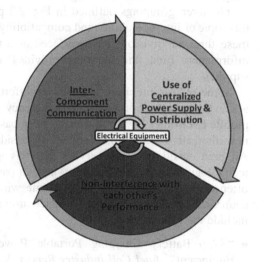

importance to manufacturers. Equipment that does not "play well" in its final application setting will likely be short-lived with the consumer.

Understandably, the concepts of interoperability and compatibility are important to equipment manufacturers. For today's fire service, the gear that they use is coming of age, and the demands of their job tasks are necessitating that mobile electronics strive for interoperability and compatibility. These will be among the deciding factors for the long-term proliferation of their equipment. Manufacturers of fire service electronic equipment need to provide gear that is interoperable and compatible with the final overall user platform, or else face evolving challenges in a competitive, consumer-driven marketplace.

For the interoperability and compatibility of any electronic equipment, the key electrical issues are rooted in the following three performance characteristics: use of centralized power supply and distribution, non-interference with each other's performance, and inter-component communication. This is illustrated in Fig. 2.2, Key Electrical Interoperability Performance Characteristics for Electronic Safety Equipment.

Each of the three characteristics identified in Fig. 2.2 is commonly addressed among every-day electronic packages. For instance, all the components with a single desktop computer are designed to work seamlessly together as a single unit, and likewise the hardware and software interfaces are designed so that multiple computers can work together.

For example, the coordination of power supplies is a base concept in the field of robotics. The use of voltage regulators and similar components are used so that equipment with different power requirements can operate together from a centralized power source. For fire fighters, there is an obvious advantage to eliminate multiple redundant power supplies (e.g., batteries) that already provide excessive weight to their overall operating ensemble (Voltage Regulator Tutorial 2012).

The three groupings outlined in Fig. 2.2 provide a convenient subdivision on this topic of interoperability and compatibility of mobile electronics. Interestingly, these three subdivisions have surfaced as a result of the exercise to collect this information. First, the literature provides multiple citations addressing power supplies.

Some of this literature is based on efforts to centralize the power source to achieve maximum portability and efficiency for the end user. Power supplies for mobile electronics have traditionally been based on various chemical batteries, but new alternative technologies are being considered, such as miniaturized fuel cells. An area outside the fire service that has given considerable attention to this technological issue is the military and, in particular, for ground soldiers who are often deployed for long periods of time with minimal ability to recharge power sources. Examples of literature oriented around enhancements of the power source include the following:

- "3-Up Battery Charging Portable Power System for Military Electronic Equipment", *Fuel Cell Industry Report*, Volume 11 (6), 2010, pg. 1. Summary of a novel portable power supply approach that maximizes efficiency for soldiers on remote duty.
- Abdelsalam, H.S., "Energy Efficient Workforce Selection in Special-Purpose Wireless Sensor Networks," Proceedings of IEEE INFOCOM Workshops, 2008. Development of an approach to efficiently task multiple electronic sensors for optimum consumption of battery power.
- Bush, S., "Mini Fuel Cell Lasts 72 Hrs," *Electronics Weekly*, Volume 2210, 2005, pg. 7. Review of a novel mini fuel cell designed for mobile electronic safety equipment.
- Fuel Cell Today Industry Review 2011, Johnson Matthey PLC, July 2011, website: http://www.fuelcelltoday.com/media/1351623/industry_review_2011.pdf, cited: 8 Feb 2011, cited: 6 February 2012. Overview of the use of fuel cell technology for applications that include the use of mobile electronics.
- Kenyon, H.S., "Soldiers' Tools Go Solar," *Signal*, Volume 61 (7), 2007, pg. 67. Utilization of solar power for the personal electronic equipment used by soldiers.

- Knoll, J.W., "A New Trend in Minified Communications Equipment," *IRE Transactions on Vehicular Communications*, Volume 9:1, 1957, pgs. 25–32. Implementation of miniaturized communications equipment specially designed for military and non-military applications.
- Mayhew-Smith, A., "Soldiers Get More Punch," *Electronics Weekly*, Vol. 2156:17, 2004. Update on efforts to maximize the electrical power supply used by soldiers.
- "Micro Fuel Processing System Can Power Today's Electronic Soldier," *Science Letter*, Volume 19, 2003. Review of the use of a novel micro fuel processing system that provides power supply for the electronic safety equipment used by ground soldiers.

Next, another convenient grouping of the literature addresses the independent functionality of the equipment in relation to other equipment in the package. Stated in another way, this is the ability of the particular equipment to function properly and not cause undue and unnecessary interference with other electronic equipment in the overall package. Some examples of this literature include the following:

- Baber, C., D.J. Haniff, and R. Buckley, "Wearable Information Appliances for the Emergency Services: HotHelmet," *Lecture Notes in Computer Science*, Volume 1707, 1999, pgs. 314–316. The development of a novel device referred to as a HotHelmet that processes wearable information.
- Bonfiglio, A., et al., "Managing Catastrophic Events by Wearable Mobile Systems," *Lecture Notes in Computer Science*, Volume 4458, 2007, pgs. 95–105. Overview of the management of an emergency incident based on the input from a wearable mobile system.
- Booher, H.R., *Handbook of Human Systems Integration*, Hoboken, N.J.: John Wiley & Sons, 2003. Basic concepts on human systems integration and its relationship to electronic safety equipment.
- Fischer, C., and H. Gellersen, "Location and Navigation Support for Emergency Responders: A Survey," *IEEE Pervasive Computing*, Volume 9 (1), 2010, pgs. 38–47. Assessment of field information relating to fire fighter location and tracking needs.
- *Improving Disaster Management: The Role of IT in Mitigation, Preparedness, Response, and Recovery*, National Academies Press, 2007. An overview of the role of information technology to enhance disaster management.
- Jiang, X., N.Y. Chen, J.I. Hong, K. Wang, L. Takayama, and J.A. Landay, "Siren: Context-Aware Computing for Firefighting," Proceedings of Pervasive Computing, April 2004, pgs. 18–23. Development of a novel approach to improve fire fighter awareness.
- Jiang, X., J. Hong, L. Takayama, and J. Landay, "Ubiquitous Computing for Firefighters: Field Studies and Prototypes of Large Displays for Incident Command," *Proceedings of ACM CHI Conference on Human Factors in Computing Systems*, April 2004, pgs. 670–686. Use of methods and support

tools to effectively process incident command information from multiple fire-ground sources.

- Leitner, G., D. Ahlström, and M. Hitz, "Usability of Mobile Computing in Emergency Response Systems—Lessons Learned and Future Directions," *Lecture Notes in Computer Science*, Volume 4799, 2007, pgs. 241–254. Update on how mobile computing is being used to benefit emergency responders.
- Moayeri, N., J. Mapar, S. Tompkins, and K. Pahlavan, "Emerging Opportunities for Localization and Tracking," *IEEE Wireless Communications*, Volume 18:2, Apr 2011, pgs. 8–9. Overview of the performance characteristics of the multiple technological approaches under consideration for indoor localization and tracking.
- Tsao, Y.C., L.C Chen, and S.C. Chan, "The Research of Using Image-Transformation to the Conceptual Design of Wearable Product with Flexible Display," *Lecture Notes in Computer Science*, Volume 4551, 2007, pgs. 1220–1229. Overview of research to develop and implement technology involving augmented reality.
- Wang, J.J.H., J.K. Tillery, K.E. Bohannan, and G.T. Thompson, "Helmet Mounted Smart Array Antenna," IEEE International Symposium on Antennas and Propagation, Volume 1, 1997, pgs. 410–413. Review of antenna design for maximum efficiency of wireless mobile electronic safety equipment.

Finally, the third of these three groupings is centered on the ability of the equipment to communicate between and among the various components. Much of this relates to wireless technologies, which today have become a mainstream technology with numerous applications throughout society. Examples identified in the literature that relate to the communicating ability of separate electronic safety equipment include the following:

- Annapurna, D., D. Shreyas Bhagavath, V. Gnanaskandan, K.B. Raja, K.R. Venugopal, and L.M. Patnaik, "Performance Comparison of AODV, AOMDV and DSDV for Fire Fighters Application," *Communications in Computer and Information Science*, Volume 250, Part 1, 2011, pgs. 363–367. Review of a wireless sensor network that maximizes the interaction between nodes to effectively elevate fire fighter communications in situations involving extreme distress.
- Desourdis, R.I, *Achieving Interoperability in Critical IT and Communication Systems*, Boston: Artech House Inc., 2009. Review of the interoperability requirements for critical electronic equipment.
- Dogra, S., S. Manna, A. Banik, S. Maiti, and S.K. Sarkar, "A Novel Approach for RFID Based Fire Protection," International Conference on Emerging Trends in Electronic and Photonic Devices & Systems, 2009, pgs. 198–201. Update on a novel approach for using RFID inter-component communication.
- Hartung, C., R. Han, C. Seielstad, and S. Holbrook, "FireWxNet: A Multi-Tiered Portable Wireless System for Monitoring Weather Conditions in Wildland Fire Environments," Paper presented at Proceedings of the 4th

International Conference on Mobile Systems, Applications and Services, 2006. Update of a novel approach to update weather for wildland fire fighters.

- Lee, E., S. Park, J. Lee, S. Oh, and S. Kim, "Novel Service Protocol for Supporting Remote and Mobile Users in Wireless Sensor Networks with Multiple Static Sinks," *Wireless Networks*, Volume 17, Number 4, 2007, pgs. 861–875. Review of a novel service protocol for mobile users of electronic safety equipment.
- Park, A.S., and S.S. Kim, "A Novel Agent-Based User-Network Communication Model in Wireless Sensor Networks," Networking '07, Proceedings of the 6th International IFIP-TC6 Conference on Ad-hoc and Sensor Networks, Wireless Networks, Next Generation Internet, 2007. Overview of a novel communications model for a wireless sensor network.
- Ruta, M., E. DiSciascio, and F. Scioscia, "RFID-Enhanced Ubiquitous Knowledge Bases: Framework and Approach," *Unique Radio Innovation for the 21st Century*, Part 3, 2010, pgs. 229–255. Summary of a framework for an RFID based approach.
- Wilson, J., V. Bhargava, A. Redfern, and P. Wright, "A Wireless Sensor Network and Incident Command Interface for Urban Firefighting," Fourth Annual International Conference on Mobile and Ubiquitous Systems: Networking and Services, Aug 2007. Review of a wireless sensor network specifically intended for fireground operations involving structural fire fighting.
- "Wireless System for Monitoring Weather Conditions in Wildland Fire Environments," Paper presented at Proceedings of the 4th International Conference on Mobile Systems, Applications and Services, 2006. Overview of a wireless system in use for wildland fire fighters to monitor weather conditions.

2.3 General Concepts for Emergency Responder Equipment

The previous section focused on interoperability and compatibility of mobile electronics, which is meant to be applicable to all mobile electronics for personal use. This section provides a refinement of these concepts by providing a more direct focus on the specific issues confronting fire fighters and other emergency responders.

A number of citations in the literature provide a general overview of certain aspects of the electronic safety equipment used by emergency responders. Examples include the following:

- Donnelly, T., "Building Collapse Rescue Operations: Technical Search Capabilities," *Fire Engineering*, Volume 163 (10), 2010, pg. 22–26. Focus on the needs of emergency responders involved with technical rescue operations such as a building collapse.

- Durso, F., "Sending and Receiving," *NFPA Journal*, Volume 105 (7), 2011, pg. 11. Overview of efforts to address fire fighter electronic safety equipment.
- Grant, C.C., "Respiratory Exposure Study for Fire Fighters and Other Emergency Responders," Fire Protection Research Foundation, Quincy, MA, December 2007. Summary of electronic equipment used by fire fighters to monitor respiratory threats and hazardous atmospheres.
- Klimenko, S.V., V. Stanislav, et al., "Using Virtual Environment Systems During the Emergency Prevention, Preparedness, Response and Recovery Phases," *NATO Science for Peace and Security Series C: Environmental Security*, Volume 6, 2008, pgs. 475–490. Use of virtual environment techniques for prevention, preparedness, response and recovery.
- Sapateiro, C., P. Antunes, G. Zurita, R. Vogt, and N. Baloian, "Evaluating a Mobile Emergency Response System," *Lecture Notes in Computer Science*, Volume 5411, 2008, pgs. 121–134. Overview of a mobile emergency response system.
- Teele, B.W., "Fire and Emergency Services Protective Clothing and Protective Equipment," *Fire Protection Handbook*, 20th edition, Section 12, Chap. 9, National Fire Protection Association, Quincy, MA, 2008, pgs 12-143–12-160. General overview information on fire fighter personal protective equipment and electronic safety equipment.

Additional literature citations provide more detailed information on specific aspects of electronic safety equipment of interest in this study. This includes the following examples:

- Bryant, R.A., K.M Butler, R.L. Vettori,, and P.S. Greenberg, "Real-Time Particulate Monitoring—Detecting Respiratory Threats for First Responders: Workshop Proceedings", *NIST Special Publication* 1051, National Institute of standards and Technology, Gaithersburg MD, December 2007. Development of electronic equipment for detecting real-time respiratory hazards.
- Donnelly, M.K., W.D. Davis, J.R. Lawson, and M.J. Selepak, *Thermal Environment for Electronic Equipment Used by First Responders*, NIST Technical Note 1474, National Institute of standards and Technology, Gaithersburg MD, January 2006. Review of the thermal insult upon all electronic safety equipment used by fire fighters.
- Kantor, G., S. Singh, R. Peterson, D. Rus, A. Das, V. Kumar, G. Pereira, and J. Spletzer, "Distributed Search and Rescue with Robot and Sensor Teams," *Springer Tracts in Advanced Robotics*, Volume 24, 2006, pgs. 529–538. Overview of concepts for using robotics and sensors for search and rescue.
- Lymberis, A., "Advanced Wearable Sensors and Systems Enabling Personal Applications," *Lecture Notes in Electrical Engineering*, Volume 75, 2010, pgs. 237–257. Review of wearable sensors for personal applications in adverse environments.
- Molino, L.N., "Electronic Control Devices and EMS," *Fire Engineering*, Volume 161 (8), 2008, pg. 30. Use of electronic safety equipment for EMS applications.

- Moore, L.K., "Emergency Communications," Nova Publishers, 2007. Overview of voice communications for emergency responders.
- Murphy, R.R., S. Tadokoro, D. Nardi, A. Jacoff, P. Fiorini, H. Choset, and A.M. Erkmen, "Search and Rescue Robotics," Springer Handbook of Robotics, Part F, 2008, pgs. 1151–1173. Update on robotics for use by emergency responders performing search and rescue tasks.
- Roberts, M.R., "NIST Tests Firefighter Tracking Devices for Radio-Frequency Interference," *Urgent Communications*, 2011. Review of a wireless sensor system by fire fighters in structural fireground conditions.
- Sun, Y., and K.Y. Ong, *Detection Technologies for Chemical Warfare Agents and Toxic Vapors*, CRC Press, 2005. Assessment of personal detection technologies used by emergency responders for hazardous environments.
- Sunderman, L.M., "Improving Firefighter Accountability Systems with the Use of Electronic Devices," Executive Fire Officer Paper, U.S. Fire Administration National Fire Academy, Emmitsburg, MD, June 2006. Using electronic systems to automatically update incident commander accountability information.
- Wei, L., Z. Zhang, Q. Wang, and A. Feng, "The Research and Implementation Based on Electronic Gases Simulation System of Wireless Sensor Network," *Advanced Materials Research*, Volume 433: 440, Jan 2012, pgs. 2519–2522. Review of wireless sensor networks based on electronic gas simulation.

The literature review reveals concepts and approaches used by other related professions. This information is helpful to draw parallels with other established work. Fire fighters depend on their electronic equipment to function and survive within a hostile environment (e.g., a room on fire). This has similarities with other professions involving applications such as space, underwater, upper atmosphere, or the battlefield. The following are some examples:

- "3-Up Battery Charging Portable Power System for Military Electronic Equipment," *Fuel Cell Industry Report*, Volume 11 (6), 2010, pg. 1. Summary of a novel portable power supply approach that maximizes efficiency for soldiers on remote duty.
- "Avionics Fly to Fire Fighters Rescue," *Machine Design*, Volume 65 (20), 1993, pg. 16. Use of concepts used for aviation pilots for adaptation to fire fighters.
- Dawson, D., "Fast Fielding for Soldier Systems," *Soldiers,* Volume 59 (3), 2004, pg. 42. Overview of concepts used for interoperability of electronic equipment in the military.
- Fei, D.Y., X. Zhao, C. Boanca, E. Hughes, O. Bai, R. Merrell, and A. Rafiq, "A Biomedical Sensor System for Real-Time Monitoring of Astronauts' Physiological Parameters During Extra-Vehicular Activities," *Computers in Biology and Medicine*, Volume 40 (7), 2010, pgs. 635–642. Use of a physiological monitoring system for astronauts on spacewalks.
- Hopmeier, M., H. Christen, and M. Malone, "Development of Human Factors Engineering Requirements for Fire Fighting Protective Equipment," Army Research Development and Engineering Command, Natick Soldier Center,

Unconventional Concepts Inc., 2005. Adapting human factors engineering techniques used by the military to the needs of fire fighters.

- House, T.B., and R.E. Strunk, "Army Soldier Enhancement Program," *Army Sustainment*, Volume 43 (1), 2011, pg. 20. An overview of efforts to enhance the efficiency and effectiveness of a typical ground soldier.
- Keggler, J., "The Soldier as Nucleus," *Armada International*, Volume 34 (4), 2010, pg. 42. Update on efforts to improve a soldier's operational effectiveness.
- Kenyon,H.S., "Soldiers' Tools Go Solar," *Signal*, Volume 61 (7), 2007, pg. 67. Utilization of solar power for the personal electronic equipment used by soldiers.
- Knoll, J.W., "A New Trend in Minified Communications Equipment," *IRE Transactions on Vehicular Communications*, Volume 9:1, 1957, pgs. 25–32. Implementation of miniaturized communications equipment specially designed for military and non-military applications.
- Kozloski, L.D., "U.S. Space Gear: Outfitting the Astronaut," Smithsonian Institution Press, 1994. Overview of the electronic gear used by astronauts.
- Mayhew-Smith, A., "Soldiers Get More Punch," *Electronics Weekly,* Vol. 2156:17, 2004. Update on efforts to maximize the electrical power supply used by soldiers.
- Möller, P., R. Loewens, I.P. Abramov, and E.A. Albats, "EVA Suit 2000: A Joint European/Russian Space Suit Design," *Acta Astronautica*, Volume 36 (1), 1995, pgs. 53–63. Details on the design of spacesuits and the electronic gear they use.
- National Research Council, Panel on Human Factors in the Design of Tactical Display Systems for the Individual Soldier, "Tactical Display for Soldiers: Human Factors Considerations," National Academies Press, 1997. Summary of the tactical display issues for ground soldiers.
- Skoog, A., "The EVA Space Suit Development in Europe," *Acta Astronautica*, Volume 32 (1), 1994, pgs. 25–38. Summary of efforts to design personal protective equipment used by astronauts on spacewalks.
- Wilde, R.C., J.W. McBarron, S.A. Manatt, H.J. McMann, and R.K. Fullerton, "One Hundred US EVAs: A Perspective on Spacewalks," *Acta Astronautica*, Volume 51 (1), 2002, pgs. 579–590. Review of experience in protecting astronauts on spacewalks.

Another aspect of the literature on emergency responder equipment concepts is that which addresses physiological monitoring. Measuring the physiological activity of an individual is well within the domain of current technology. However, the bigger challenge is how to then process that information. Some of this literature explores how an individual fire fighter or incident commander could react based on the complex real-time information that could be collected through physiological monitoring. The following are some examples:

- Barr, D., T. Reilly, and W. Gregson, "The Impact of Different Cooling Modalities on the Physiological Responses in Firefighters During Strenuous Work Performed in High Environmental Temperatures," *European Journal of*

Applied Physiology, Volume 111 (6), 2011, pgs. 959–967. Review of the physiological stress experienced by fire fighters in high temperature environments.

- Fei, D.Y., X. Zhao, C. Boanca, E. Hughes, O. Bai, R. Merrell, and A. Rafiq, "A Biomedical Sensor System for Real-Time Monitoring of Astronauts' Physiological Parameters During Extra-Vehicular Activities," *Computers in Biology and Medicine*, Volume 40 (7), 2010, pgs. 635–642. Use of a physiological monitoring system for astronauts on spacewalks.
- Mordecai, M., "Physiological Status Monitoring for Firefighters," *Firehouse*, Volume 33 (9), 2008, pg. 112. Update on efforts to monitor the physiological of fire fighters.
- Pallauf, J., P. Gomes, S. Bras, J.P. Cunha, and M. Coimbra, "Associating ECG Features with Firefighter's Activities," *Annual Conference Proceedings of the Conference of the IEEE Engineering in Medicine and Biology Society*, 6009–12, August 2011. Measurement of the levels of stress in a sample of fire fighters while performing actual job tasks based on their electrocardiogram signals.
- Plog, B.A., and P.J. Quinlan, "Fundamentals of Industrial Hygiene," National Safety Council, 5th edition, 2002. Basic information on fundamental parameters for maintaining a realistic working environment.
- Whitby, C., "Firefighter Ergonomics Enhance Equipment Performance," *Fire Engineering*, Volume 163 (6), 2010, pg. 101. Focus on approaches to provide efficient and effective ergonomics of personal protective equipment used by fire fighters.

2.4 Specific Emergency Responder Equipment

There are a variety of published items in the literature that address specific electronic safety equipment used by emergency responders. These are examined here based on the general category of equipment.

A review of the specific equipment used by today's fire fighter suggests that literature can be grouped into five primary categories. These are: (a) communications; (b) environmental monitoring; (c) physiological monitoring; (d) sensory support; and (e) tracking/location. This is illustrated in Fig. 2.3, Basic Categories of Emergency Responder Electronic Safety Equipment.

The traditional method for fire fighters to communicate with each other on the fireground is through portable radios that transmit and receive voice communication. However, it's acknowledged that aside from the communication of actual speech, all the electronic equipment will ultimately need to communicate if it is to function transparently in a full interoperable and compatible electronic platform.

This concept is represented by the shaded bar within the bottom of Fig. 2.3 that represents the additional equipment to equipment communication, and which is applicable to all mobile equipment and thus addressed in the preceding section.

Fig. 2.3 Basic categories of emergency responder electronic safety equipment

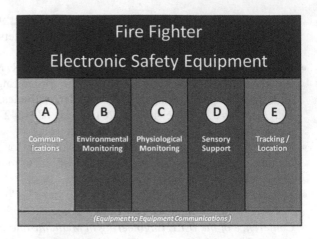

Examples of citations in the literature relating to communications, and specifically voice communications, include the following:

- Davis, W.D., M.K. Donnelly, M.J. Selepak, and N. Building, "Testing of Portable Radios in a Fire Fighting Environment," NISTIR 1477, National Institute of Standards and Technology, Building and Fire Research Laboratory, 2006. Review of functionality testing of radios in environments typically experienced by fire fighters.
- Faulhaber, G.R., "Solving the Interoperability Problem: Are We on the Same Channel? An Essay on the Problems and Prospects for Public Safety Radio," *Federal Communications Law Journal*, Volume 59 (3), 2007, pg. 493. Update on efforts to address interoperability of radio equipment.
- Fatah, A.A., et al., "Guide for the Selection of Communication Equipment for Emergency First Responders," *NIJ Guide 104–00*, National Institute of Justice, Washington DC, Volume I, Feb 2002. Overview guidance to select radio equipment used by emergency responders for fireground conditions.
- Frazier, P., R. Hooper, and B. Orgen, "Current Status, Knowledge Gaps, and Research Needs Pertaining to Firefighter Radio Communication Systems," NIOSH, Morgantown, WV, and TriData Corporation, Arlington, VA, Sept 2003. Update of current status, knowledge gaps, and research needs pertaining to radio communications used by fire fighters.
- Goldstein, H., "Radio Contact in High-Rises can Quit on Firefighters," *IEEE Spectrum*, Volume 39 (4), 2002, pgs. 24–27. Review of challenges of portable radio operation within high-rise buildings.
- Merrill, D., "Cranking Up the Heat on Firefighters' Radios," *Mobile Radio Technology*, Volume 25 (7), 2007, pg. 4. Review of need to enhance and improve radio technology used on the fireground.
- Spahn, E.J., *Fire Service Radio Communications*, Fire Engineering, NY, 1989. Overview of the use of radio technology by fire fighters during fireground operations.

- Tuite, D., "Radio Interoperability: It's Harder Than It Looks," *Electronic Design,* Volume 56 (8), 2008, pg. 30. Review of the challenges to address radio equipment interoperability.

Several citations can be found in the literature relating to the monitoring of the hazardous environment that fire fighters are likely to encounter. These generally take the form of gas analyzers, temperature measuring devices, etc. Some examples of literature relating to this topic include the following:

- Bryant, R.A., K.M. Butler, R.L. Vettori, and P.S. Greenberg, "Real-Time Particulate Monitoring—Detecting Respiratory Threats for First Responders: Workshop Proceedings," *NIST Special Publication* 1051, National Institute of standards and Technology, Gaithersburg, MD, December 2007. Development of electronic equipment for detecting real-time respiratory hazards.
- Chou, J., *Hazardous Gas Monitors: A Practical Guide to Selection, Operation and Applications*, McGraw-Hill Book Company, NY, 2000. Review of guidance for selecting gas analyzers for use in hazardous atmospheres.
- Fatah, A.A., et al., "An Introduction to Biological Agent Detection Equipment for Emergency First Responders," *NIJ Guide 101–00*, National Institute of Justice, Washington DC, December 2001. Overview of information on gas analyzers specifically intended for emergency events involving biological agents.
- Grant, C.C., "Respiratory Exposure Study for Fire Fighters and Other Emergency Responders," Fire Protection Research Foundation, Quincy, MA, December 2007. Summary of electronic equipment used by fire fighters to monitor respiratory threats and hazardous atmospheres.
- Hawley, C., *Hazardous Materials Air Monitoring and Detection Devices*, Thomas Delmar Learning, 2nd edition, 2007. Basic information on the technology used for monitoring the environment found in fireground conditions.
- Sun, Y., and K.Y. Ong, *Detection Technologies for Chemical Warfare Agents and Toxic Vapors*, CRC Press, 2005. Assessment of personal detection technologies used by emergency responders for hazardous environments.
- Wei, L., Z. Zhang, Q. Wang, and A. Feng, "The Research and Implementation Based on Electronic Gases Simulation System of Wireless Sensor Network," *Advanced Materials Research*, Volume 433:440, Jan 2012, pgs. 2519–2522. Review of wireless sensor networks based on electronic gas simulation.

The monitoring of the physiological activity is also a category of use that is receiving significant attention. Fire fighting is, at times, an extremely stressful and demanding occupation, and it is valuable for individual fire fighters to be aware of their physiological conditions in real time. Further, this is valuable information for incident commanders, and it allows oversight for the personnel on the fireground. For certain other applications, this technology is already well developed, with remote physiological monitoring applied to astronauts and soldiers, as well as in non-hostile environments such as candidates attending drafts for professional sports (Zephyr 2012).

Here, Personal Alert Safety Systems (PASS) are included in this grouping based on their intent to monitor the overall physiological status of a fire fighter. Arguably this could also appear in other categories such as tracking and location. Applicable citations in the literature include the following:

- Bryner, N., D. Madrzykowski, and D. Stroup, "Performance of Thermal Exposure Sensors in Personal Alert Safety System (PASS) Devices," NISTIR 7295, National Institute of Standards and Technology, Sep 2005. Assessment of thermal impact on the specific components used in PASS.
- Buyan, M., P. Bruhwiler, A. Azens, G. Gustavsson, R. Karmhag, and C. Granqvist, "Facial Warming and Tinted Helmet Visors," *International Journal of Industrial Ergonomics*, Volume 36:1, Jan 2006, pgs. 11–16. Review of thermal impact to SCBA face pieces and its physiological consequences.
- Mazza, S.L., "An Evaluation of Self Contained Breathing Apparatus Voice Communication Systems," EFO Paper, United States Fire Administration, May 2008. An evaluation of the manual voice communication feature normally used in conjunction with SCBA.
- McNamee, R.B., "The Use of Personal Alert Safety Devices to Decrease Levels of Firefighter Risk of Death and Injury," EFO Paper, United States Fire Administration, Dec 1994. Review of the value of PASS.
- "PASS Signals Can Fail at High Temps," *Fire Chief*, IAFF, Volume 49 (12): 10, 2005, pg 10. Update on the challenges of thermal impact on the proper operation of PASS.

Certain electronic safety equipment is used to directly enhance the sensory inputs already available to humans. The handheld flashlight is within this category and is perhaps the most commonly found piece of electronic equipment on the fireground. Other equipment of interest includes thermal imaging cameras, video recording cameras for training or documentation, and similar equipment. Certain equipment is being addressed in other sections based on other factors, like PASS, which is based on audible locator-based technology. Literature of interest relating to electronic safety equipment that is related to sensory support includes the following:

- Amon, F., N. Bryner, and A. Hamins, "Thermal Imaging Research Needs for First Responders: Workshop Proceedings," NIST Special Publication 1040, National Institute of Standards and Technology, Jun 2005. Documentation of a workshop that focused on identifying the needs of emergency responders on the use of thermal imaging cameras.
- Amon, F., A. Hamins, N. Bryner, and J. Rowe, "Meaningful Performance Evaluation Conditions for Fire Service Thermal Imaging Cameras," *Fire Safety Journal*, Volume 43:8, Nov 2008, pgs. 541–550. Review of performance characteristics for thermal imaging cameras.
- Amon, F.K., D. Leber, and N. Paulter, "Objective Evaluation of Imager Performance," Fifth International Conference on Sensing Technology, Dec 2011,

pgs. 47–52. Update on efforts to establish performance parameters for thermal imaging cameras.

- Mandal, R., and I. Singh, "Optical Design and Salient Features of an Objective for a Firefighting Camera," *Optical Engineering*, Volume 46, Aug 2007. Overview of basic design features for thermal imaging cameras.
- Richardson, M., and R. Scholer, "Thermal Imaging Training: Covering the Basics," *Firehouse*, Vol. 26, No. 4, 2001, pp. 86–88. Review of basic training issues relating to thermal imaging cameras.
- Wilson, J., D. Steingart, R. Romero, J. Reynolds, E. Mellers, A. Redfern, L. Lim, W. Watts, C. Patton, J. Baker, and P. Wright, "Design of Monocular Head-Mounted Displays for Increased Indoor Firefighting Safety and Efficiency," *Proceedings of SPIE*, Volume 5800, 2005. Update on a novel approach to providing hands-free information for fire fighters on the fireground.

In the last several years there has been considerable effort to enable advanced firefighter locator technology. Advanced locator technology has some noteworthy distinctions from traditional audible PASS technology and is considered here separately.

For example, audible PASS technology is currently in widespread use as a simple, technologically mature, and relatively dependable last-resort mechanism for locating fire fighters needing immediate assistance. Advanced locator technology is in its infancy and has yet to overcome significant technological hurdles that are preventing its widespread field application. In the meantime, existing PASS technology is the established backbone of locating fire fighters needing immediate rescue.

If and when advanced locator technology overcomes its technological challenges, the established use of audible PASS devices is not expected to be replaced, but more likely it will be supplemented. Additional applicable reports, articles, and other information in the literature from the perspective of alternative locator and tracking technology include the following:

- Bonfiglio, A., et al., "Managing Catastrophic Events by Wearable Mobile Systems," *Lecture Notes in Computer Science*, Volume 4458, 2007, pgs. 95–105. Overview of the general technology used in wearable mobile systems.
- Dogra, S., S. Manna, A. Banik, S. Maiti, and S.K. Sarkar, "A Novel Approach for RFID Based Fire Protection," International Conference on Emerging Trends in Electronic and Photonic Devices and Systems, 2009, pgs. 198–201. Assessment of a novel approach for RFID tracking.
- Duckworth, J.R., "Tracking Lost Firefighters: Firefighter Search & Rescue Systems Demonstrated at WPI Workshop," *Firehouse Magazine*, Volume 35 (10), 2010, pg. 94. Review of workshop to clarify efforts to advance tracking technology for the fire service.
- Fischer, C., and H. Gellersen, "Location and Navigation Support for Emergency Responders: A Survey," *IEEE Pervasive Computing*, Volume 9 (1), 2010, pgs. 38–47. A review of survey information on efforts to advance locator technology for fire fighters.

- Foster, S., "GPS System is Lifeline to Firefighters," *Computing Canada*, Volume 30 (4), 2004, pg. 16. Update on the need to advance locator technology.
- Klann, M., "Tactical Navigation Support for Firefighters: The LifeNet Ad-Hoc Sensor-Network and Wearable System," *Lecture Notes in Computer Science*, Volume 5424, 2009, pgs. 41–56. Assessment of a specific approach to implement a proposed network wearable tracking system.
- Miller, L.E., "Indoor Navigation for First Responders: A Feasibility Study," National Institute of Standards and Technology, Advanced Network Technologies Division, 10 Feb 2006, website: https://www.hsdl.org/?view&did=478117, cited: 6 Feb 2012. Feasibility study on the inherent technological challenges of indoor navigation for fire fighters.
- Moayeri, N., J. Mapar, S. Tompkins, and K. Pahlavan, "Emerging Opportunities for Localization and Tracking," *IEEE Wireless Communications*, Volume 18:2, Apr 2011, pgs. 8–9. Overview of efforts to advance tracking and locator technology for the fire service.
- Putorti Jr, A.D., F.K. Amon, K.M. Butler, C.A. Remley, W.F. Young, and C. Spoons, "Structural and Electromagnetic Scenarios for Firefighter Locator Tracking Systems," NIST TN 1713, National Institute of Standards and Technology, Gaithersburg, MD, 2011. Addresses locator communication challenges of attenuation and in-band and out-of-band interference and multipath for fire fighter locators and other communication devices.
- Ramirez, L., T. Dyrks, J. Gerwinski, M. Betz, M. Scholz, and V. Wulf, "Landmarke: an Ad Hoc Deployable Ubicomp Infrastructure to Support Indoor Navigation," *Personal and Ubiquitous Computing*, Volume 10, 2007. Review of a proposed novel approach to support indoor navigation technology for fire fighters.
- Rantakokko, J., J. Rydell, P. Stromback, P. Handel, J. Calmer, D. Tornqvist, F. Gufstafsson, M. Jobs, and M. Gruden, "Accurate and Reliable Soldier and First Responder Indoor Positioning: Multisensor Systems and Cooperative Localization," *IEEE Wireless Communications*, Volume 18:2, Apr 2011, pgs. 10–18. An update of tracking and locator technology being used by the military.
- Roberts, M.R., "NIST Tests Firefighter Tracking Devices for Radio-Frequency Interference," *Urgent Communications*, 2011. Assessment of possible interference between proposed tracking and locator technology and traditional voice communication radios.

2.5 Regulations, Model Documents and Policy

Various citations within the literature address regulatory and/or policy issues, which directly relate to the electronic safety equipment used by emergency responders. Several of these literature items focus on applicable high level guidance, typically oriented in public policy. Examples include the following:

- Committee on Homeland and National Security, Subcommittee on Standards, "A National Strategy for CBRNE Standards," *National Science and Technology Council,* May 2011. A national strategy for the U.S. federal government for an approach to standardization for terrorist events involving CBRNE (chemical, biological, radiological, nuclear, or explosives).
- Committee on Using Information Technology to Enhance Disaster Management, "Summary of a Workshop on Using Information Technology to Enhance Disaster Management," National Academies Press, 2005. Documentation of a workshop addressing information technology for disaster management, hosted by the U.S federal government.
- Krimgold, F., K. Critchlow, and N. Uda-gama, "Emergency Service in Homeland Security," NATO Security Through Science Series, 2006, pgs. 193–229. Role of emergency services based on NATO activities.

Various regulatory-based documents are available that address or relate to the topic of electronic safety equipment of emergency responders. These are generally in the form of consensus-developed model codes and standards. Examples include the following:

- ANSI/ISA-S82.03, *Safety Standard for Electrical and Electronic Test, Measuring, Controlling and Related Equipment,* 1988. Model standard developed through the consensus-based process of Instrument Society of America, addressing test methods for electronic equipment.
- FM 3610, *Approval Standard for Intrinsically Safe Apparatus and Associated Apparatus for Use in Class I, II, and III, Divisions 1, Hazardous (Classified) Locations,* January 2010. Model standard developed by FM Approvals, addressing the intrinsic safety of electronic safety equipment in hazardous atmospheres.
- MIL-PRF-28800F, Performance Specification, "General Specification for Test Equipment with Electrical and Electronic Equipment," 24 June 1996, website: http://www.everyspec.com/MIL-PRF/MIL-PRF+(010000+-+29999)/MIL-PRF-28800F_18207/, cited: 30 January 2012. Performance specification used by the U.S. Department of Defense for the testing of electronic safety equipment.
- NFPA 1801, *Standard on Thermal Imagers for the Fire Service,* National Fire Protection Association, Quincy, MA, 2010 edition. Model standard developed through the consensus-based process of National Fire Protection Association, directly addressing thermal imaging cameras used by emergency responders.
- NFPA 1981, *Standard on Open-Circuit Self-Contained Breathing Apparatus (SCBA) for Emergency Services,* National Fire Protection Association, Quincy, MA, 2007 edition. Model standard developed through the consensus-based process of National Fire Protection Association, directly addressing SCBA and associated electronics used by emergency responders.
- NFPA 1982, *Standard on Personal Alert Safety Systems (PASS),* National Fire Protection Association, Quincy, MA, 2007 edition. Model standard developed through the consensus-based process of National Fire Protection Association, directly addressing PASS used by emergency responders.

- *Proposed Standard on Electronic Safety Equipment for Emergency Services,* Pre-ROP Draft, National Fire Protection Association, Quincy, MA, 17 Nov 2006. Proposed model standard developed through the consensus-based process of National Fire Protection Association, directly addressing electronic safety equipment used by emergency responders.
- UL 913, *Intrinsically Safe Apparatus and Associated Apparatus for Use in Class I, II, and III, Division 1, Hazardous (Classified) Locations,* Underwriters Laboratories, Northbrook IL, 31 July 2012. Model standard developed through the consensus-based process of Underwriters Laboratories, addressing the intrinsic safety of electronic safety equipment in hazardous atmospheres.

Of the various regulatory documents that relate to electronic safety equipment, one model document is of particular interest. This is the *Proposed Standard on Electronic Safety Equipment for Emergency Services.* This document was started in the early- to mid-2000 time frame, but was tabled due to the responsible technical committee focusing their attention on other more pressing document development initiatives.

This proposed standard directly addresses the key elements of interest in this study on electronic safety equipment used by emergency responders. Because of the importance of this draft standard, the information it contains serves as a useful backdrop to the information being collected in this study.

2.6 Future Trends

Predicting future trends is always a challenge, and this is true with electronics in general and portable electronic safety equipment used by emergency responders in particular. The advancements of new technology in today's world are often noteworthy and occasionally dazzling.

Understandably, manufacturers and others are continually exploring new technological applications through their own research and development programs, sometimes in conjunction with external centers of knowledge such as academia. These lead to the fruits of a competitive marketplace and are based on the inherent incentive to provide superior marketplace products that meet or exceed the optimum performance characteristics of fire service users. Clearly, portable ESE that is interoperable and compatible is a desirable holistic feature that would increase the likelihood of its ultimate success in the marketplace.

It is typical for such work to be done by manufacturers and others on a proprietary basis until fully released, with the intent of maximizing development ahead of competition prior to product launching. Thus, examples in the literature are more challenging to locate for equipment still in development phases. However, examples do exist. One such example is a new futuristic helmet, referred to as "C-thru," with multiple features built into it as a comprehensive package for fire fighters. (FireRescue1Staff, "Futuristic New Helmet Helps Firefighters See Through Smoke," 2012) In this case the product is being developed in Sweden, and in addition to the conventional helmet performance features provided such as impact and heat protection, the unit integrates multiple electronic technologies such as an optical thermal camera, head-mounted projection display, cloud computing, target acquisition, and selective active noise cancellation

One useful summary of emerging technologies represented by products under development can be found in the "In Development Technologies" section of the Responder Knowledge Base (RKB) website. (USFEMA, "In Development Technologies," Responder Knowledge Base) This provides a continually updated summary of information on a variety of emergency responder related technologies that are in development, and it includes federally-funded technologies as well as general market research for new commercial technologies. The mission of the RKB is to provide emergency responders, purchasers, and planners with a trusted, integrated, online source of information on products, standards, certifications, grants, and other equipment-related information.

Certain programs are underway that might involve multiple industry partners and are being promoted through government funding to address a perceived need. A good example of such an effort is an ongoing research project supported by the U.S. Department of Homeland Security referred to as GLANSER, which stands for Geospatial Location Accountability and Navigation System for Emergency Responders. (USDHS, "Where There's Smoke, There's a Signal," July 2011) This program is attempting to overcome the inherent technical challenges of tracking fire fighters within structures, which is difficult to do with the necessary accuracy and precision due to the components of modern building construction that result in significant interference. GLANSER utilizes microwave radio technology and a suite of navigation devices to transmit individual fire fighter location back to the on-scene incident command.

A review of the literature reveals that there are certain technological directions that are showing promise. These are reviewed here for purposes of considering future advances with portable electronic safety equipment. Further elaboration of the technical subject-areas that are summarized in Table 2.3, Examples of Concept Areas for Future Technological Development, will follow in the subsequent section.

Table 2.3 Examples of concept areas for future technological development

a. Augmented reality

b. Electronic textiles and physiological monitoring

c. Next generation power supplies

d. Wireless advanced infrastructure

2.6.1 Augmented Reality

One promising concept based on a different application of available technology is augmented reality. This involves the display of computer information as an overlay on top of the user's field of vision, without the use of a fixed terminal or other traditional interactive display system. Fire fighting in hostile environments appears to be an ideal application for this technological approach.

An augmented reality system would allow fire fighters with head-mounted displays to interact with a computer using a virtual interface that is continually available in a non-disruptive manner, such as on the palm of one glove. This technology also introduces significant opportunities for training by allowing three-dimensional representation of virtual environments, both for individual fire fighters in hostile environments and on a larger scale for incident commanders managing the fireground. Examples of references and citations addressing this topic include:

- Bretschneider, N., S. Brattke, and R. Karlheinz, "Head Mounted Display for Fire Fighters," 3rd International Forum on Applied Wearable Computing, Mar 2006, pgs. 1–15. Overview of augmented reality used in head-mounted displays for fireground applications.
- Klimenko, S.V., V. Stanislav, et al., "Using Virtual Environment Systems During the Emergency Prevention, Preparedness, Response and Recovery Phases", *NATO Science for Peace and Security Series C: Environmental Security*, Volume 6, 2008, pgs. 475–490. Review of augmented reality based systems for emergency responder training applications.
- Steingart, D., J. Wilson, A. Redfern, P. Wright, R. Romero, and L. Lim, "Augmented Cognition for Fire Emergency Response: An Iterative User Study," University of California Berkeley, paper presented at Proceedings of Augmented Cognition, HCI International Conference, 2005. Assessment of an information technology network involving head mounted displays and an infrastructure to gather and deliver information to firefighters (Steingart 2005).
- Tanagram, "About Us," website: http://tanagram.com/2012/01/24/defining-augmented-reality-and-why-it-will-be-a-performance-enhancing-technology/, cited: 24 January 2012. Review of a novel application of head-mounted display augmented reality (HMD-AR) specifically designed for fire fighters in a hostile environment.

2.6.2 Electronic Textiles and Physiological Monitoring

Another novel application of technology relating to electronic safety equipment is electronic textiles, also known as e-textiles. This is based on microtechnology and nanotechnology that is introducing electronic wearable garments, with electronic capabilities built in the garment materials. They would function like other electronic safety equipment, with the ability to locally and remotely monitor environmental conditions, physiological conditions, tracking/location, and so on. Examples of literature addressing this topic include the following:

- Bonfiglio, A., et al., "Managing Catastrophic Events by Wearable Mobile Systems," *Lecture Notes in Computer Science*, Volume 4458, 2007, pgs. 95–105. Review of electronic textiles used in a system for fireground applications.
- Jayaraman, S., P. Kiekens, and A.M. Grancaric, "Intelligent Textiles for Personal Protection and Safety," NATO Public Diplomacy Division, NATO Programme for Security through Science, IOS Press, Volume 3, 2006. Review of electronic textiles for personal protection of NATO representative.
- Kohler, A.R., L.M. Hilty, and C. Bakker, "Prospective Impacts of Electronic Textiles on Recycling and Disposal," *Journal of Industrial Ecology*, Volume 15:4, Aug 2011, pgs. 496–511. Assessment of recycling and disposal of discarded and retired electronic textiles based on their assumed proliferation.
- Kowalski, K.M., "Electronics Textiles Wiring the Fabrics of our Lives," *Odyssey*, Volume 15 (6), 2006, pg. 13. Overview of the applications involving electronic textiles.
- Marculescu, D., R. Marculescu, Z. Stanley-Marbell, K. Park, J. Jung, L. Weber, et al., "Electronic textiles: A Platform for Pervasive Computing", *Proceedings of the IEEE*, Volume 91 (12)2003, pgs. 1993–1994. Review of the use of electronic textiles in networking applications.

One high-profile program of particular interest that is directly addressing the physiological monitoring of fire fighters is PHASER, which stands for Physiological Health Assessment System for Emergency Responders. (USDHS, "Where There's Smoke, There's a Signal," July 2011) This ongoing research project is independent of electronic textiles and is supported by the U.S. Department of Homeland Security, operating in conjunction with their GLANSER research activity. PHASER involves the development of equipment that monitors a fire fighter's body temperature, blood pressure, and pulse, and transmits these details back to the fireground incident commander.

2.6.3 Next Generation Power Supplies

Power supplies are an essential part of electronic safety equipment working on a common unified platform. This was discussed earlier in this literature review, based on the need to improve ergonomic efficiency by eliminating redundant power supplies. Yet it is mentioned again here because novel technological innovations are being developed to alter the power supplies for mobile electronics altogether. Different advanced chemical battery arrangements are being developed as enhancements to traditional nickel-cadmium and lithium-ion batteries. Full replacements to chemical batteries are also looming on the horizon, such as solar powered or fuel cell arrangements, and micro-furnace applications. Some of the literature on this includes the following:

- Bush, S., "Mini Fuel Cell Lasts 72 Hrs," *Electronics Weekly*, Volume 2210, 2005, pg. 7. Review of a novel mini fuel cell designed for mobile electronic safety equipment.
- Fuel Cell Today Industry Review 2011, Johnson Matthey PLC, July 2011, website: http://www.fuelcelltoday.com/media/1351623/industry_review_2011.pdf, cited: 8 Feb 2011, cited: 6 February 2012. Overview of the use of fuel cell technology for applications that include the use of mobile electronics.
- Kenyon, H.S., "Soldiers' Tools Go Solar," *Signal*, Volume 61 (7), 2007, pg. 67. Utilization of solar power for the personal electronic equipment used by soldiers.
- "Micro Fuel Processing System Can Power Today's Electronic Soldier," *Science Letter*, Volume 19, 2003. Review of the use of a novel micro fuel processing system that provides power supply for the electronic safety equipment used by ground soldiers.

2.6.4 Wireless Advanced Infrastructure

The topic of inter-communications was discussed earlier and involves the full implementation of wireless networks and infrastructure. The future direction of this technological application is suggesting that the wireless network of tomorrow could possibly contain much more than the mobile electronic safety equipment we think of today (Klimenko 2008).

The availability of data in today's world is increasing at rates that were unimaginable a few short years ago. Mobile sensors and communications devices are being embedded with increasing regularity. The use of private cell phones, for example, has proliferated at amazing rates throughout society.

This proliferation of sensing and communication technology extends to equipment and appliances as well as to people. Everything from automobiles to small hand tools are being equipped with the ability to communicate and function within the bigger environment within which it's placed. A strikingly simple example are electronic door chocks used to hold open doors, which are equipped with location tracking and operational status features to assist with operational use and inventory control (Delaney 2009).

Handling the increasing, massive amounts of data being collected is a daunting challenge, but it's one that's already the subject of focus for technologists and innovators. A concept that is gaining widespread recognition is referred to as "Big Data," which acknowledges our capture of massive amounts of data that can then be used to tell the big picture story (Manyika 2011).

Stepping back from the macro applications of the "Big Data" concept, such as using it to predict global economic trends, we can apply this just as readily in a more limited sense to situations such as a complex fireground. It is not beyond our comprehension for the incident commanders of tomorrow to have immediate real-time data on the location and operational status of all fire fighting personnel, equipment, and even the structure itself under their command on a fire scene (Werner 2000).

Chapter 3
PPE Electronic Inventory

The PPE electronic landscape is expansive and continually evolving. A general discussion is provided on this landscape, with particular emphasis on the Responder Knowledge Base that provides the most up-to-date summary of the available products within the subject area.

3.1 Responder Knowledge Base

Today the most useful summary of equipment used by emergency responders is the responder knowledge base (RKB). (USFEMA, "In Development Technologies," Responder Knowledge Base) The RKB is a web-based summary of product related information administered by the Federal Emergency Management Agency of the U.S. Department of Homeland Security (FEMA/USDHS). The RKB is unique and provides a single comprehensive summary of product information for emergency responders.

The RKB provides a continually-updated summary of information on a variety of emergency responder–related technologies that are in development, and it includes federally-funded technologies as well as general market research for new commercial technologies. The mission of the RKB is to provide emergency responders, purchasers, and planners with a trusted, integrated, online source of information on products, standards, certifications, grants, and other equipment-related information.

The RKB is relatively robust, and this is partly due to it addressing all emergency responders, with the fire service as but one target audience among others (e.g., law enforcement and EMS). Further, the RKB addresses all types of fire fighters, with specific focus on fire service personnel other than structural and proximity, such as those who are engaged in wildland fire fighting or in technical rescue.

The RKB is web-based and thus continually undergoing revisions and updates. The RKB database has evolved based on two separate lists. The first of these two

C. C. Grant, *Interoperable Electronic Safety Equipment*,
SpringerBriefs in Fire, DOI: 10.1007/978-1-4614-8277-2_3,
© Fire Protection Research Foundation 2012

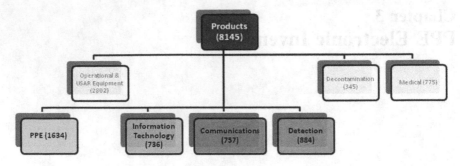

Fig. 3.1 Responder knowledge base—product overview

lists is the *Authorized Equipment List* (AEL), whose full title is the FEMA Preparedness Grants Authorized List. The AEL is generated by the FEMA Grant Programs Directorate under the U.S. Department of Homeland Security, and it provides the official generic list of equipment items allowable under several DHS grant programs, including the Homeland Security Grant Program.

The second primary list in the RKB database is the *Standardized Equipment List* (SEL), and the full title is the Inter Agency Board Standardized Equipment List. This list is generated by the Inter Agency Board for Equipment Standardization and Interoperability (IAB), and it contains the minimum equipment recommendations for planned response to incidents involving weapons of mass destruction (WMD).

The RKB database includes both the AEL and SEL as separate lists, but conveniently it also provides an integrated list that consolidates both lists. This is referred to as the *Integrated AEL/SEL Display*, and it displays the information in a single record for items that likewise appear on both the AEL and SEL.

To exemplify the information provided by the RKB, the details of different products are illustrated in Fig. 3.1. This is representative of the type of information included in the RKB, which in the RKB "Product Summary" includes 8,145 products as of the date this information was accessed (July 2012). The products are further subdivided into seven subcategories, and these are shown in Fig. 3.1. The number shown in parenthesis within each block represents the number of products included in the Product Summary list.

Portable ESE tends to be included under most categories within the RKB Product Summary, but it is particularly applicable with regard to the following four subcategories: PPE; Information Technology; Communications; and Detection. These are the four subcategories that are highlighted in Fig. 3.1.

An overview of products summarized in the "PPE" subsection of the RKB is illustrated in Fig. 3.2. This represents 1,634 different pieces of PPE-related equipment.

A summary of the products in the "Information Technology" subsection of the RKB is illustrated in Fig. 3.3. This represents 736 different pieces of equipment related to Information Technology.

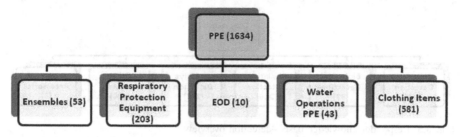

Fig. 3.2 Responder knowledge base—personal protective equipment overview

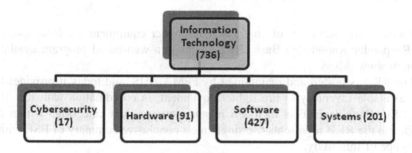

Fig. 3.3 Responder knowledge base—information technology overview

A summary of the products in the "Communications" subsection of the RKB is illustrated in Fig. 3.4. This represents 757 different pieces of equipment related to Communications

A summary of the products in the "Detection" subsection of the RKB is illustrated in Fig. 3.5. This represents 884 different pieces of Detection-related equipment.

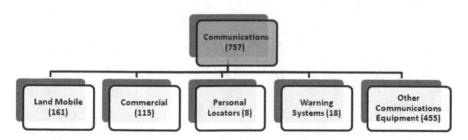

Fig. 3.4 Responder knowledge base—communications overview

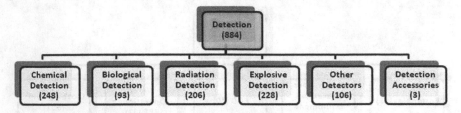

Fig. 3.5 Responder knowledge base—detection overview

3.2 Inventory Summary Observations

The most useful summary of emergency responder equipment available today is the Responder Knowledge Base (RKB). This is a web-based program available through www.rkb.us.

The RKB is hosted and maintained by FEMA/DHS, and today it provides the best available inventory of fire fighter equipment, in coordination with the AEL and SEL. Portable ESE is included in numerous product categories within the RKB, and the RKB represents the single most conclusive summary of ESE within the scope of this study.

Chapter 4
Performance Requirements for Interoperability

This study provides support to the emergency response community by promoting the concept of a standardized platform/framework for electronic safety equipment. Previous sections of this report have provided a conceptual overview of this subject area based on a review of the literature and have presented an understanding of the inventory of existing and emerging electronic equipment categorized by key area of interest to the fire service.

This section will address equipment performance requirements relevant to interoperability, which as used herein is understood to include concepts of integration and compatibility. Performance requirements address important features such as communications, power requirements, and intrinsic safety, and they support an action plan for the development of requirements to meet the needs of emergency responders. Standardization within the ESE marketplace provides the primary basis for an action plan to promote the concepts of interoperability.

4.1 Definitions

A key aspect for embracing concepts in support of ESE performance requirements are the definitions of certain key terms. Perhaps foremost among important terms is the definition of ESE, Electronic Safety Equipment.

The genesis of the specific use of the term ESE in the NFPA standards process can be traced back to the approval by the NFPA Standards Council in April of 2002 of the NFPA Technical Committee on Electronic Safety Equipment. While not rigidly defined at that time, the general understanding of the topic pertains to electronic equipment that is used by fire fighters and other emergency responders as part of their personal protective equipment.

Once the Technical Committee on Electronic Safety Equipment was established and committee members appointed, they initiated work on a *Proposed Standard on Electronic Safety Equipment for Emergency Services* (Proposed Standard 2006). Through this venue, a definition of ESE evolved as follows: "Products that contain

C. C. Grant, *Interoperable Electronic Safety Equipment*,
SpringerBriefs in Fire, DOI: 10.1007/978-1-4614-8277-2_4,
© Fire Protection Research Foundation 2012

3.3.6 <u>Electronic Safety Equipment (ESE)</u>. Products that contain electronics embedded in or associated with the product for use by emergency services personnel that provides enhanced safety functions for emergency services personnel and victims during emergency incident operations.

from Proposed Standard on Electronic Safety Equipment, section 3.3.6, 2006

A review of the full range of equipment used by fire fighters supports this definition, although the following feature requires further clarification:

Question: Is this intended to include or exclude **Portable, Mobile, Stationary** and/or **Field Deployable** Equipment?

Fig. 4.1 Clarifying the definition of "ESE"

electronics embedded in or associated with the product for use by emergency services personnel that provides enhanced safety functions for emergency services personnel and victims during emergency incident operations." This is illustrated in Fig. 4.1.

As this overall topic evolves, a question has arisen as to the portability of ESE. It is generally understood that this is intended for equipment that is worn by an emergency responder as they work within the fireground environment. However, this is not clearly stated, and questions linger as to whether or not this is intended to include equipment that may or may not be used while attached directly to operational personnel. Is it meant for equipment that is portable, mobile, stationary, and/or field deployable? This technical detail still requires further clarification and should be addressed by the Technical Committee on Electronic Safety Equipment.

Another term that requires clarification of its definition is "interoperability." This term is already used within the emergency response community in a broad sense to address the necessary interactions between wireless communications (i.e., portable radios). As such, a definition of "ESE Interoperability" is offered to clarify the distinction with wireless communication. This is illustrated in Fig. 4.2.

The term "interoperability" can have a different meaning depending on the sphere of communication within which it is used. This is particularly noteworthy with regard to wireless communication on the fireground and as represented by Fig. 4.3.

The four spheres of fireground communication in Fig. 4.3 are as follows: personal area network; team/unit; fireground incident command; and inter-jurisdiction. The differences in communication methods and protocols in each sphere are typically distinct. For example, the wireless communication between ESE components on a fire fighter have different interoperability concerns than the fireground communications between units or the incident commander, and with inter-jurisdictional communications involving mutual aid.

In general, **interoperability** refers to the ability of emergency responders to work seamlessly with other systems or products without any special effort.

Wireless Communications Interoperability specifically refers to the ability of emergency response officials to share information via voice and data signals on demand, in real time, when needed, and as authorized.

from DHS SAFECOM project (see www.safecomprogram.gov)

Interoperability is the ability to operate in synergy in the execution of assigned tasks.

from Dictionary of Military & Associated Terms (see www.definitions.net/definition/interoperability)

ESE Interoperability is the ability of ESE to operate in synergy in the execution of assigned tasks.

Fig. 4.2 Clarifying the definition of "Interoperability"

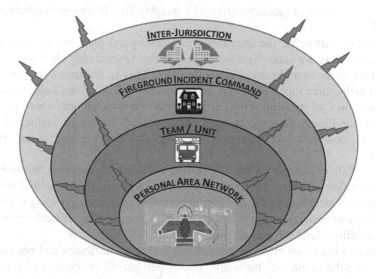

Fig. 4.3 Overview of fireground wireless communications

4.2 Interoperability Concepts

A key thrust of this research study is to clarify concepts involving interoperability and promote these concepts as appropriate. Earlier, important ESE performance characteristics were illustrated in Fig. 2.2, Key Electrical Interoperability Performance Characteristics for ESE. This had a specific focus toward the interoperability of *electrical* components. Now, we are elevating this concept beyond only

Fig. 4.4 Interoperability conceptual platform—key performance characteristics

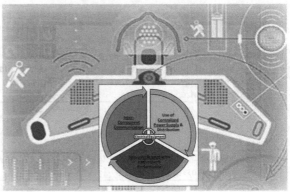

the electrical aspects of the components to consider additional important parameters such as physical characteristics like weight. This is shown symbolically in Fig. 4.4.

Expanding further on the interoperability conceptual platform (as indicated in Figs. 2.2 and 4.4), we naturally want to expand this beyond only electrical and also account for other non-electrical characteristics. This is illustrated in Fig. 4.5. In addition to the three focus areas that are electrically oriented (i.e., inter-component communication, centralized power supply, and non-interference), there are other pertinent non-electrical performance characteristics. Examples include the following: ergonomics; user interface; donning and doffing; and form, fit, and function.

The overall evaluation of the component attributes is important to assess the wholesale value added of the ESE in question. These are represented in Fig. 4.6, which address interoperability and compatibility as primary supporting features of operability and with operability itself as an equal, among other critical attributes of maintainability, durability, availability, stability, and reliability.

There is a continual effort to maximize all of these attributes and not sacrifice one for the sake of others. One single, poorly-performing attribute is all it takes to derail an otherwise promising ESE product. For example, a product that is not available due to cost or manufacturing limitations, or is a maintenance nightmare, will cripple an ESE component no matter how promising its other attributes may be.

Another performance characteristic of interest is how the ESE is applied or delivered on the fireground. This is illustrated in Fig. 4.7. Earlier, the basic categories of ESE were separated in Fig. 2.3 into the five logical categories that follow: (a) communications; (b) environmental monitoring; (c) physiological monitoring; (d) sensory support; and (e) tracking/location. These are embedded within Fig. 4.7.

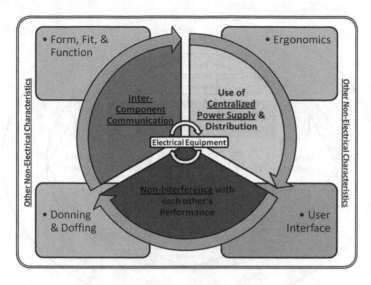

Fig. 4.5 Key interoperability performance characteristics for ESE

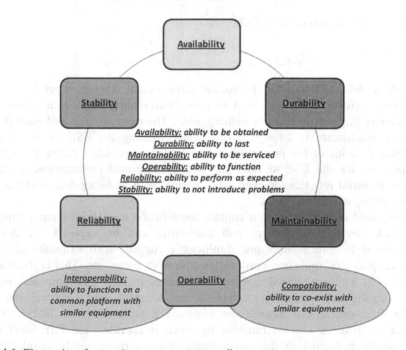

Fig. 4.6 Electronic safety equipment component attributes

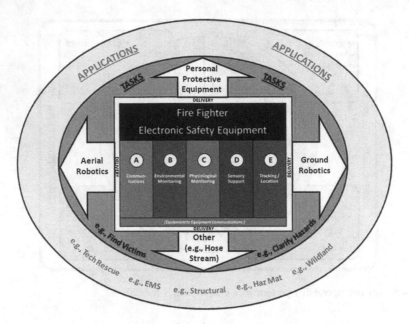

Fig. 4.7 Application and delivery of ESE

ESE is delivered into the fireground environment through several possible means to perform tasks (e.g., find victims, clarify hazards, etc.) in support of applications (e.g., structural fire fighting, etc.). The most obvious and anticipated delivery mechanism is through a fire fighter wearing the ESE as an integral component on his or her protective ensemble. However, other delivery methods are possible for the ESE to enter the hostile fireground environment, such as ground or aerial robotics, or even other means, such as being delivered via the extinguishing agent.

Fireground environments can contain some of the most hostile environments in which electrical equipment and electronics can be exposed. Conditions encountered by fire fighters are significantly rugged, with extremes of temperature, pressure, moisture, particulates, corrosive gases, etc. This is illustrated in Fig. 4.8. The anticipated operational hazards include environmental influences, conditional exposure, mechanical impingement, and electrical impact. While these operational hazards are generally a factor in terms of how they impact the ESE in the environment in which it operates, the ESE itself can likewise be a hazard to the environment (e.g., intrinsic safety of electrical equipment).

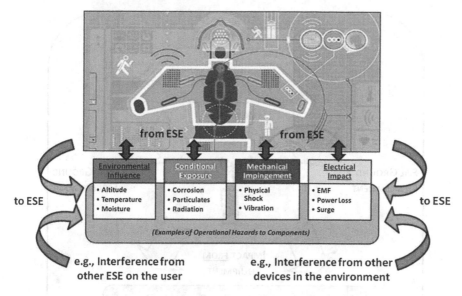

Fig. 4.8 Interoperability conceptual platform—primary ESE operational hazards

4.3 Portable Equipment Intrinsic Safety

Fire fighters often operate in hazardous environments that include gases, dusts, and other dangerous substances. When operating in these environments, caution must be taken to minimize and control possible ignition sources. This includes the portable ESE that fire fighters carry with them as they operate in these environments.

In the fireground environment, care is required for the *impact* **to** *equipment* from environmental influence, conditional exposure, mechanical impingement, and electrical impact. Just as important, however, care is also required for the *impact* **from** *equipment*, so that the ESE does not itself create an additional hazard. This concept is illustrated in Fig. 4.9.

A key approach taken with all electrical equipment is to design it so that it will not provide an ignition source while in a hazardous environment. Intrinsic safety is one of several protection techniques for electrical equipment in hazardous environments. There are different levels of intrinsic safety that can be designed into equipment, and in general, the more intrinsically safe the design, the more other features may be compromised with trade-offs such as portability, weight, size, cost, power supply, etc.

The traditional designation of hazardous environments in North America is based on requirements in *NFPA 70®, National Electrical Code®*, which uses a system of three classes, two divisions, and multiple groups and temperature

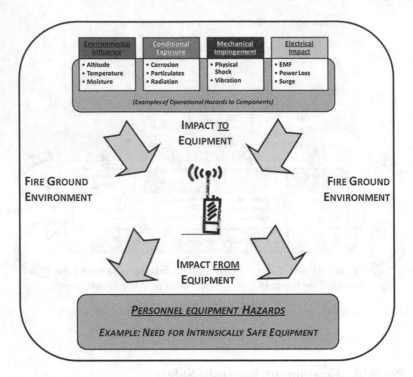

Fig. 4.9 Overview of fireground intrinsic safety

Table 4.1 Overview of hazardous environments

	Class I		Class II		Class III	
Hazardous substance	Combustible gases		Combustible dust		Combustible filings	
	Division 1	*Division 2*	*Division 1*	*Division 2*	*Division 1*	*Division 2*
Hazardous presence	Normal	Abnormal	Normal	Abnormal	Normal	Abnormal
	Group		*Group*		*Group*	
Damage potential	(A) Acetylene		(E) Combustible metal dusts		None	
	(B) Hydrogen equivalents		(F) Carbonaceous dusts			
	(C) Ethylene and similar		(G) Grain, wood, plastic dusts			
	(D) Acetone, propane, etc.					

classes. Table 4.1, Overview of Hazardous Environments, provides an overview of how hazardous environments are categorized and identified (Intrinsic Safety and Thermal Imaging 2012). This approach is being further harmonized with the zone classification system used in Europe and other regions.

4.4 Interoperability Standardization

Traditional research studies that collect and analyze data provide an overall value to the mainstream body of knowledge, but they are limited in their ability to drive change. Implementing meaningful change requires coordinating the broad constituent base, and the most effective means for doing so is through standards. Here, standardization represents the primary basis for an action plan to promote the concepts of interoperability.

Technical safety standards represent the will of society on complex technological subjects. Although we often strive for a perfect world, this is not possible as there will always be some degree of risk. Thus, our standardization efforts strive to balance risk and/or performance with the resources we have available to us. This is represented in Fig. 4.10.

Standards provide a fundamental means for transferring technology from science to practice. Of particular interest are consensus-developed standards that are generated as models for application by others in a mandatory or non-mandatory setting. The North American approach for the development of model standards has evolved into a robust infrastructure that has a proven reputation for providing competent documents in a transparent and fair manner.

These documents are developed based on broad and balanced input, and whose ultimate purpose is to address complex technical subjects. The best mechanism for implementing enhancements to fire fighter safety and health is through consensus-developed standards that are practically based, adequately validated, and broadly accepted by the affected constituency.

There are two primary types of standards of interest for the topic of portable ESE. First are standards that directly address the subject and provide details on design, operation, manufacture, certification, maintenance, and other key functional characteristics. The second are standards that address portable ESE indirectly as part of a bigger program without focus on functional details.

The former are the documents of most direct interest to the user community, as this is where they are able to directly address their users' concerns for performance characteristics. This is also the primary venue for addressing issues relating to

Technical Standards:
- Represent the will of society on complex technical issues

Represent the balance between...
- Acceptable risk and/or performance
- Available resources (e.g., cost)

User Perspective

Fig. 4.10 The balance of technical standardization

interoperability. Examples of this latter type are standards that require or otherwise enable the use of portable ESE, but they do not directly address the details of the subject. A specific example is chapter 7 of NFPA 1500, *Standard on Fire Department Occupational Safety and Health Program*, which provides general requirements for when and where to use ESE (NFPA 1500 2007).

Of paramount interest are standards that directly address portable ESE, and a key model standard is the *Proposed Standard on Electronic Safety Equipment for Emergency Services* (Proposed Standard 2006). This draft document was developed by the NFPA Technical Committee on Electronic Safety Equipment. They started working on this initiative in the early- to mid-2000 time frame, but then tabled their efforts due to other commitments.

Table 4.2 Document outline for proposed standard on electronic safety equipment for emergency services

Chapter 1 *Administration*
Chapter 2 *Referenced publications*
Chapter 3 *Definitions*
Chapter 4 *Certification*
Chapter 5 *Product labeling and information*
Chapter 6 *Design requirements*
    6.1 General design requirements
    6.2 Electronics classification
    6.3 Design safety for electronic safety equipment (ESE)
    6.4 ESE systems design approach and criteria
    6.5 Transmission quality
Chapter 7 *Performance requirements*
    7.1 ESE for Hostile fire environments performance requirements
    7.2 ESE for Hostile non-fire environments performance requirements
Chapter 8 *Test methods*
    8.1 Sample preparation
    8.2 Function recognition test
    8.3 Signal frequency test
    8.4 Environmental temperature stress test
    8.5 Immersion/leakage test
    8.6 Vibration test
    8.7 Case integrity test
    8.8 Impact acceleration resistance test
    8.9 Corrosion test
    8.10 Viewing surface and lens abrasion test
    8.11 Heat resistance test
    8.12 Heat and flame test
    8.13 Tumble—vibration test
    8.14 Product label durability test
    8.15 Cable pullout test
Annex A: *Explanatory material*
Annex B: *Explanatory material*

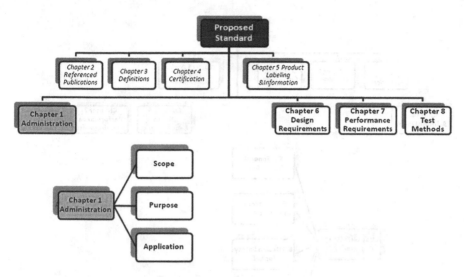

Fig. 4.11 Proposed standardization framework—chapter 1 administration

The proposed standard directly addresses the key elements of interest in this study. The information it contains serves as a useful backdrop to the information being collected. However, this proposed standard has evolved based on individual ESE devices and would need re-alignment to promote interoperability concepts. This is further clarified by Table 4.2, Document Outline for Proposed Standard on Electronic Safety Equipment for Emergency Services (Proposed Standard 2006).

A significant amount of work was put into the development of this draft document. This will lend itself well as a leverage point when standardization efforts are revisited. The basic framework and many of the details are still generally applicable, and interested parties would benefit from salvaging as much of this material as possible, with necessary and appropriate re-alignment.

In particular, the administrative chapter of the proposed standard can be reviewed and re-utilized for a renewed standardization effort, with a focus on updating the scope, purpose, and application. This is illustrated in Fig. 4.11.

The importance of reconfirming the document's scope, purpose, and application should not be underestimated, as this sets the foundation for what is and what is not to be included within the standardization initiative. Other portions of the proposed standard can be reused with the appropriate perfunctory updating, such as referenced publications, definitions, certification, and product labeling information.

Additional important criteria are addressed via design requirements, and this was included in the previous chapter 6 of the earlier proposed standard. This is illustrated in Fig. 4.12. Here, the design requirements should be re-evaluated based on the previous subcategories of general requirements, electronic classification, design safety for ESE, ESE system design approach, and transmission quality.

The performance requirements of ESE are another important area of consideration. This is addressed in Fig. 4.13.

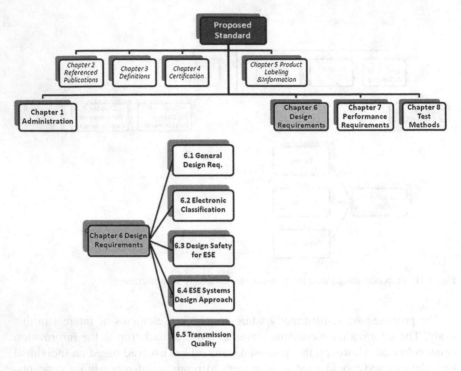

Fig. 4.12 Proposed standardization framework—chapter 6 design requirements

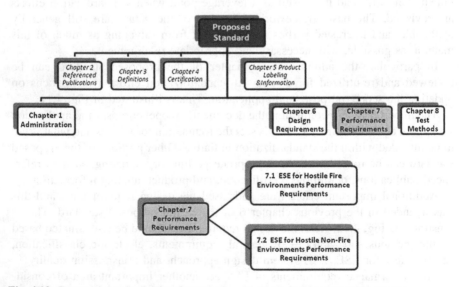

Fig. 4.13 Proposed standardization framework—chapter 7 performance requirements

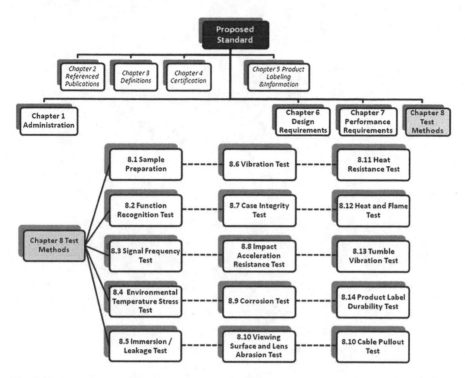

Fig. 4.14 Proposed standardization framework—chapter 8 test methods

In the earlier proposed standard this was addressed in chapter 7 and subdivided the requirements into either "ESE for hostile fire performance requirements," or "ESE for hostile non-fire environments performance requirements." This needs to be reconsidered to account for all possible scenarios and situations where ESE is expected to perform. Further, this is the appropriate venue to address concepts of integration, compatibility, and interoperability, and these concepts should be promoted for the full ESE package on the host fire fighter.

The one additional section that should be addressed consists of the specific test methods to be used to qualify the ESE. This is represented in Fig. 4.14. In addition to the methods needed to prepare the test specimen, the following tests should be considered: function recognition; signal frequency; environmental temperature stress; immersion/leakage; vibration; case integrity; impact acceleration resistance; corrosion; viewing surface and lens abrasion; heat resistance; heat and flame; tumble—vibration; product label durability; and cable pullout.

Chapter 5
PPE Electronic Equipment Workshops

Two separate workshops were held in conjunction with this project and in support of project deliverables. The first was held on 19 September 2012 and addressed the specific sub-issue of intrinsic safety of portable electronic equipment used by emergency responders in hazardous environments. The second was held on 20 September 2012 and addressed the central deliverables of this project and focused on the performance requirements for emergency responder interoperable and compatible electronic safety equipment.

5.1 Intrinsic Safety of Portable Electronic Safety Equipment

A workshop titled "Intrinsic Safety of Emergency Responder Electronic Safety Equipment" was held on Wednesday 19 September 2012 at Underwriters Laboratories in Northbrook Illinois. This event is also referred to herein by the short title of "ESE Intrinsic Safety Workshop," as well as simply the "Intrinsic Safety Workshop".

The goal of this workshop was to clarify recommended parameters and levels of intrinsic safety for ESE in hostile fire ground environments, with a particular focus on portable communication radios. This is a sub-issue of the broader topic of interoperability and compatibility of ESE, and ultimately supports the specific objectives of this project. The agenda for the workshop is summarized in Table 5.1, Intrinsic Safety Workshop Agenda.

As background, ESE is in widespread use by fire fighters and other emergency responders, and it provides essential support for performance of specific job task. Often this equipment is required to perform in hostile environments, and as a result it is generally required to be intrinsically safe for use in combustible and flammable atmospheres. The genesis of this topic is rooted in portable communication radios, though it applies to all emergency responder ESE used in hazardous environments.

C. C. Grant, *Interoperable Electronic Safety Equipment*,
SpringerBriefs in Fire, DOI: 10.1007/978-1-4614-8277-2_5,
© Fire Protection Research Foundation 2012

Table 5.1 Intrinsic safety workshop agenda

	Welcome and introductions	10:00 am
TBD	(1) Review of workshop goal, objectives, and deliverables	10:05 am
TBD (UL)	(2) Presentation on ESE intrinsic safety standardization	10:25 am
TBD (FM)	(3) Presentation on ESE intrinsic safety standardization	10:45 am
TBD	(4) Case Study presentation of current and future radio ESE	11:05 pm
TBD	(5) Presentation of fire service ESE performance requirements	11:25 am
TBD	(6) Review and preliminary discussion of information presented	11:45 pm
	Networking lunch	12:00 pm
All Participants	(7) Discussion of intrinsic safety problem	1:00 pm
All Participants	(8) Analysis and prioritization of intrinsic safety needs	2:00 pm
All Participants	(9) Development of guiding principles and recommended action steps	3:00 pm
	Adjourn	4:00 pm

Multiple levels of intrinsic safety are generally defined in various standards arenas, and a question exists as to the appropriate level of intrinsic safety for portable ESE required to operate in adverse atmospheres. Portable communication radios are being used in this workshop as a specific case study focus, based on lack of clarity for proposed standard requirements addressing the design of ESE that promote higher levels of intrinsic safety but potentially compromise other user performance characteristics.

The constituent groups interested in and involved with the workshop were from the four basic areas that follow: (1) fire service users; (2) enabling standards; (3) intrinsic safety product standards; and (4) radio manufacturers. These are visually represented in Fig. 5.1. The fire service users represent all aspects of those professional individuals and organizations that ultimately use the equipment, including those who purchase, operate, service, maintain, and train with the equipment. The enabling standards tend to be model documents from NFPA and elsewhere that provide no detail on intrinsic safety requirements but instead require it as a performance characteristic of a particular device. The intrinsic safety product standards are those model standards that evaluate the specific intrinsic safety features of a particular device (e.g., FM & UL). The radio manufacturers produce the equipment in the marketplace, and this is intended to include those organizations that also approve or certify the equipment.

The list of stakeholders that participated in the workshop is summarized in Table 5.2, Attendees at the ESE Intrinsic Safety Workshop. The opening workshop remarks included the observation that opportunities for direct dialogue

Fig. 5.1 Overview of intrinsic safety workshop constituents

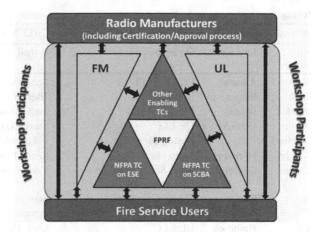

among the workshop participants in the past have been relatively limited, and this workshop is facilitating important one-on-one engagement to discuss various aspects of this topic. Participant demographics included organizational representatives and subject matter experts that are involved with fireground-related ESE, and especially portable communication radios. The constituent groups in attendance included fire service emergency responders, manufacturers, standards writing representatives, and applicable subject matter experts.

An important question for the community that designs, manufactures, and uses fire service ESE is often asked: What level of intrinsic safety should be applied to the equipment? A critical venue for establishing these requirements are through the model standards that enable these requirements (e.g., NFPA standards). These are aptly referred to as the "enabling" standards, which provided a central focus of discussion at the workshop.

Of particular interest are three enabling standards that have widespread recognition among the NFPA family of model consensus fire service standards. This is shown in Fig. 5.2, Examples of Fire Service Enabling Standards. The fire service user community depends on NFPA standards that enable the use of ESE by providing the appropriate performance requirements. In this case, three existing NFPA standards address specific ESE as follows:

(1) NFPA 1801, *Standard on Thermal Imagers for the Fire Service*
(2) NFPA 1981, *Standard on Open-Circuit Self-Contained Breathing Apparatus (SCBA) for Emergency Services*
(3) NFPA 1982, *Standard on Personal Alert Safety Systems (PASS)*

An additional new NFPA standards activity was recently established to address the performance requirements for portable communications radios, though this activity is relatively new and has not yet generated a standard. Meanwhile, Fig. 5.2 provides a direct comparison of the intrinsic safety requirements for the three existing NFPA ESE enabling standards. For sake of discussion herein, the "enabling" standards are considered to be the various NFPA standards that

Table 5.2 Attendees at the ESE intrinsic safety workshop

Attendees ESE intrinsic safety workshop 19 September 2012 NorthbrookIL

First name	Last name	Organization	E-mail
Bob	Athanas	FDNY/SAFE-IR	boba@safe-ir.com
Joel	Berger	Kenwood USA	JBerger@kenwoodusa.com
Patrick	Byrne	FM Approvals	patrick.byrne@fmapprovals.com
Steve	Corrado	UL LLC	Steven.D.Corrado@ul.com
Michael	Feely	Boston Fire Dept.	mikef. bfd@citvof boston. gov
Sandy	Florence	Motorola	Sandv.Florence@motoralasoiutions.com
Craig	Gestler	Mine Safety Appliance Company	Craig.Gestler@MSANet.com
Scott	Glazer	Thales Communications	Scott.Elazer@thalescomminc.com
Casey	Grant	FPRF	cgrant@nfpa.org
Chris	Hasbrook	UL LLC	Chris.Hasbrook@ul,com
Bill	Haskell	NIOSH NPPTL	whaskell@cdc.gov
Gordon	Hong	OTTO	gordon.hong@ottoexcellence.com
Mark	Keck	Motorola Solutions	cmk012@motorolasolutions.com
Paul	Kelly	UL LLC	Paui.T.Kellv@UL.com
Jae	Kim	OTTO	Jae.Kim@ottoexcellence.com
William	Lawrence	FM Approvals	Wiiliam.Lawrence@fmapprovals.com
Mark	Lee	OTTO	Mark.Lee@ottoexcellence.com
Robert	Martell	FM Approvals	robert.martell@fmapprovals.com
Gary	McCarraher	IAFC; Franklin MA FD	ffdchief@hotmail.com
Kerry	McManama	UL LLC	Kerry.McManama@ul.com
John	Morris	ISG Infrasys	jmorris@isgfire.com
Joel	Pfeffer	Thales Communications	Joseph.Pfeffer@thalescomminc.com
Dan	Rossos	Portland Fire St Rescue (OR)	dan.rossos@portlandoregon.gov
Trey	Sato	Kenwood USA	ksato@kenwoodusa.com
Benjamin	Schaefer	UL LLC	Benjamin.Schaefer@ul.com
Robert	Speidel	Harris Corporation	rspeidel@harris.com
Greg	Spillane	Motorola Solutions	greg.spillane@motorolasolutions.com
Christina	Spoons	West Dundee Fire Dept (IL)	cspoons@msn.com
Jeff	Stull	International Personnel Protection	intlperpro@aol.com
Steven	Townsersd	City of Carrollton Fire Dept (TX)	steventownsend@verizon.net
Dave	Trebisacci	NFPA	dtrebisacci@nfpa.org
Bruce	Varner	NFPATCon ESE	Bruce@bhvarner.com
Melanie	Wieser	OTTO	Melanie.Wieser@ottoexcellence.com

provide user-oriented performance requirements for ESE, and refer to "product" standards that contain detailed intrinsic safety requirements.

With respect to the information presented in Fig. 5.2, it's noted that NFPA 1981 and NFPA 1982 are just completing their latest revision cycles for the new 2013 editions of each respective standard. Based on a review of the proposed revisions

> **NFPA 1801, STANDARD ON THERMAL IMAGERS FOR THE FIRE SERVICE (2010)**
>
> **7.1.6**
> Thermal imagers shall be tested for listing to ANSI/ISA-12.12.01, Nonincendive Electrical Equipment for Use in Class I and II, Division 2 and Class III, Divisions 1 and 2 Hazardous (Classified) Locations, and shall meet the requirements for at least Class I, Division 2, Groups C and D hazardous locations, and with a Temperature Class of T3 or T4 or T5 or T6. For the purpose of the impact test referenced in 15.4 of ANSI/ISA 12.12.01, NFPA 1801 shall be considered the applicable standard for products in unclassified locations.

> **NFPA 1981, STANDARD ON OPEN-CIRCUIT SCBA FOR EMERGENCY SERVICES (2007)**
>
> **6.1.8**
> All electric circuits integral to an SCBA or to any SCBA accessories shall be certified to the requirements for Class I, Groups C and D; Class II, Groups E, F, and G, Division 1 hazardous locations specified in ANSI/UL 913, *Standard for Intrinsically Safe Apparatus and Associated Apparatus for Use in Class I, II, and III, Division 1 Hazardous (Classified) Locations*.

> **NFPA 1982, STANDARD ON PERSONAL ALERT SAFETY SYSTEMS (PASS) (2007)**
>
> **5.1.8**
> PASS also shall be labeled as certified at least to the requirements for Class I, Groups C and D; and Class II, Groups E, F, and G; Division 1 hazardous locations specified in ANSI/UL 913, *Standard for Intrinsically Safe Apparatus and Associated Apparatus for Use in Class I, II, and III, Division 1 Hazardous (Classified) Locations*.

Fig. 5.2 Examples of fire service enabling standards

to each standard, at the time of the compilation of these proceedings no changes have been detected to paragraph 6.1.8 of NFPA 1981 or 5.1.8 of NFPA 1982 for the upcoming 2013 editions. Further, these paragraphs refer to ANSI/UL 913, and additional reference details are summarized in the reference-document chapter in each respective standard, which is chapter 2 of both NFPA 1981 and NFPA 1982. The respective chapter 2 information will provide a specific reference to the sixth edition of ANSI/UL 913, based on the following text: "ANSI/UL 913, *Standard for Intrinsically Safe Apparatus and Associated Apparatus for Use in Class I, II, and III, Division 1 Hazardous (Classified) Locations*, Sixth Edition."

With regard to intrinsic safety of electrical equipment, fire service users are asking the following critical question: What should be the stated intrinsic safety requirements in these three existing standards and in the new forthcoming standard on portable communication radios? Two important observations can be drawn from the side-by-side comparison in Fig. 5.2. These are as follows:

- First, it is noted that the requirements are different for the different ESE. Should intrinsic safety requirements be the same for all ESE, and if not, is there logical rationale for these differences?
- Second, what should be the actual intrinsic safety requirements? Should they be, for example, Division 1 or Division 2?

With this information as background, the following comments were made through the remainder of the workshop discussions:

General ESE Intrinsic Safety

- The two factors that have altered the landscape and forced ongoing revisions to the product standards addressing intrinsic safety are:

  – Advancements in technology
  – Harmonization of the international marketplace

- A mix of opinions exist on the specificity of the NFPA standards intrinsic safety requirements, with some indicating the need to provide specific citations to specific intrinsic safety standards, while others indicate the specific intrinsic safety standards should not be cited. The difference of opinions included:

  – Specific product standards addressing intrinsic safety are not all the same and the NFPA enabling standards should reference the specific product standard per normal NFPA citation approach.
  – Specific citations are not necessary and the focus should be on suitability for a specific area classification rather than a specific protection method.

- The margin of safety for ESE intrinsic safety is not the same for all equipment and has eroded over time due to technological enhancements and other factors, though it is still within acceptable parameters.
- A probabilistic assessment would be beneficial to clarify where and when fire fighters expect to operate within a hazardous environment. For example, the most likely scenario for fire fighters is that they will encounter a hazardous environment where it's not expected, such as a gas leak. In such cases, the equipment on-site (without intrinsic safety protection) would arguably introduce a more realistic danger than the fire fighters' own equipment.
- A centralized interoperable platform that combines and maximizes the efficiency of multiple power supplies would potentially also alleviate other concerns, such as the current trade-off for reduced power supplies to comply with more rigorous Division 1 requirements.

Consistency of ESE Intrinsic Safety Requirements between NFPA Standards

- A baseline approach would be to have all ESE with similar and consistent intrinsic safety requirements, and if not, then proper rationale should be provided.
- Different ESE necessitates different intrinsic safety requirements, as some devices are relatively simple with low power consumption and minimal electronic components (e.g., PASS), while others require higher levels of power and have more robust electronics (e.g., thermal imaging camera).
- In the field, all fireground ESE is used similarly, and it is hard to tactically distinguish choices in the field (i.e., which equipment should appropriately be used in which hazardous environment).
- Among the NFPA standards to address intrinsic safety, PASS was the first to do so in the 1970s/1980s, and at that time it was believed the NFPA standard

defaulted to Division 1 requirements, though the technical basis for doing so is not clear.

- A task group of the NFPA ESE committee recently provided a detailed re-evaluation of intrinsic safety requirements for PASS. (Note: following the workshop, this information was obtained and further clarified below). This provides a model that should be considered for other ESE.

Need for Division 1 or Division 2 Intrinsic Safety Requirements

- The difference between Division 1 and Division 2 environments does not mean the equipment is more reliable or durable.
- A rigorous assessment should be considered to clarify when and where fire fighters anticipate encountering hazardous environments. This would help clarify the requirements for either Division 1 or Division 2 levels of safety. These classifications are presently based on normal or abnormal probability of occurrence, but this is meant for the electrical equipment installed at a fixed location rather than portable equipment moving in and out of hazardous locations.
- Some trade-offs can compromise performance, such as a Division 1 requirement that limits portable radio power and results in poor communication performance.
- Manufacturer feedback confirms that strengthening the intrinsic safety requirements can directly and adversely affect performance for certain ESE, such as portable radios.
- It makes sense to default to Division 1 if there are no adverse trade-offs for performance. (This seems to be the case for flashlights, SCBA, and PASS, but trade-offs may need to be considered for thermal imaging cameras, radios, and certain other electronic equipment.)
- The Division 1 versus Division 2 question should be revisited regularly, since advancements in technology have a direct effect on these requirements.

Subsequent to the workshop, the aforementioned report (that was conducted by the Intrinsic Safety Task Group on behalf of the NFPA Technical Committee on Electronic Safety Equipment) was located. This reported on an evaluation of PASS equipment for Division 1 or Division 2 classification, and it focused on this topic for the previous revision cycle of NFPA 1982, *Standard on Personal Alert Safety Systems*. The report concluded that it was appropriate to continue requiring PASS to be classified according to Division 1 requirements. Figure 5.3, PASS Evaluation Summary from the Intrinsic Safety Task Group, illustrates the evaluation summary for PASS.

A key point during discussions at the workshop focused on the hazardous environments where fire fighter ESE is intended to be used. The needs of the fire service are unique relative to the exposure to hazardous environments of other intrinsically safe equipment, as the equipment is not permanently installed in a fixed location but rather portable on a fire fighter that moves in and out of environments that are questionably hazardous. To facilitate future ongoing discussion on this point, this concept is illustrated in Fig. 5.4, Overview of Fire Service Hazardous Environments.

Intrinsic Safety Task Group

Summary

1. Based on the written responses from the manufacturers, virtually all of the performance parameters can realize improvements varying from 0% to about 30% by changing from Div 1 to Div 2.
2. Verbal comments from the remaining PASS manufacturers indicated that they wished to stay at Div 1 certification level.
3. The notable exceptions are possible improvements for RF Transmit Power of 0% to 60%, and depend on which manufacturer is making the product.
4. The decrease in safety by going from Div 1 to Div 2 depends on the total number of "events," the total number of times any firefighter throughout the US enters a potentially explosive environment over NFPA revision cycle (5 years?) or life of product (10 years?). While difficult to pinpoint, this number may exceed 10M (10,000,000).
5. In order to ensure that no explosion occurs for PASS devices, as the total number of events increases, the probability of a single failure in the device must be significantly lower for Div 2-certified devices compared to Div 1-certified devices.

Fig. 5.3 PASS evaluation summary from the intrinsic safety task group

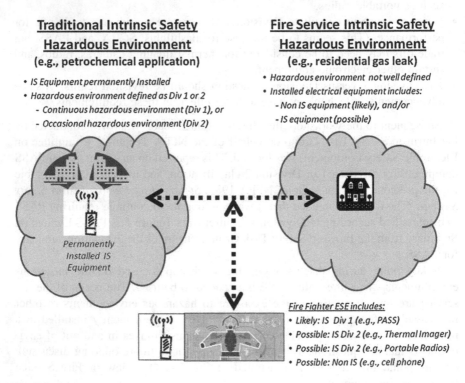

Traditional Intrinsic Safety Hazardous Environment
(e.g., petrochemical application)

- IS Equipment permanently installed
- Hazardous environment defined as Div 1 or 2
  - Continuous hazardous environment (Div 1), or
  - Occasional hazardous environment (Div 2)

Permanently Installed IS Equipment

Fire Service Intrinsic Safety Hazardous Environment
(e.g., residential gas leak)

- Hazardous environment not well defined
- Installed electrical equipment includes:
  - Non IS equipment (likely), and/or
  - IS equipment (possible)

Fire Fighter ESE includes:
- Likely: IS Div 1 (e.g., PASS)
- Possible: IS Div 2 (e.g., Thermal Imager)
- Possible: IS Div 2 (e.g., Portable Radios)
- Possible: Non IS (e.g., cell phone)

Fig. 5.4 Overview of fire service hazardous environments

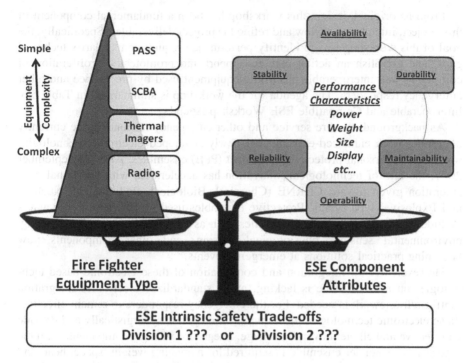

Fig. 5.5 ESE intrinsic safety trade-offs

Another key discussion point addressed the trade-offs between the type of ESE and the trade-offs for performance/attributes based on whether the equipment is designed for either a Division 1 or Division 2 environment. As ESE continues to evolve with more robust features (e.g., circuitry, display interfaces, power supply, etc.), it continues to raise questions on how the choice of intrinsic safety requirements impact the component attributes and performance characteristics.

For example, the choice of Division 1 or Division 2 for portable radios has a direct influence on the available power supplies and, in turn, directly affects the operability of the radio performance. This is conceptually illustrated in Fig. 5.5, ESE Intrinsic Safety Trade-offs.

5.2 Interoperable and Compatible Electronic Safety Equipment

A workshop titled "Performance Requirements for Emergency Responder Interoperable and Compatible Electronic Safety Equipment" was held on 20 September 2012 in Atlanta, Georgia. This workshop is also referred to herein by the short title of "Interoperable ESE Workshop" as well as "Interoperability Workshop."

From its original design, this workshop has been a fundamental component of this project, intended to review and refine the project deliverables. Specifically, the goal of this workshop was to identify performance requirements, clarify technical gaps, and establish an action plan to support and promote the proliferation of compatible and interoperable electronic equipment used by fire service and other emergency responders. The agenda for the workshop is summarized in Table 5.3, Interoperable and Compatible ESE Workshop Agenda.

As background, the fire service and other emergency responders are currently benefiting from enhanced-existing and newly-developed electronic technologies for use with personal protective equipment (PPE) ensembles. After 11 September 2001, the rate of technological innovation has accelerated, with additional consideration given toward CBRNE (Chemical, Biological, Radiological, Nuclear, and Explosive) type events. Protective ensembles used by emergency responders include or will soon include electronics such as communications, GPS/tracking, environmental sensing, physiological sensing, and other components now becoming practical solutions at emergency events.

However, overall integration and coordination of these electronic-based technologies on a broad scale is lacking, and a standardized electronics integration platform/framework is needed. For the emergency responders to remain effective, these electronic technologies must interact and operate synergistically and provide an effective and efficient overall package. Integration of these components with the emergency responder ensemble is required for managing weight, space, heat, and power requirements, as well as to create the least interference and user burden.

This workshop was held in conjunction with the NFPA Technical Committee on Electronic Safety Equipment, which is the NFPA committee responsible for the *Proposed Standard on Electronic Safety Equipment for Emergency Services*. This document is directly related to the deliverables of this project, and support of this initiative and/or similar initiatives are important for the proliferation of the project outcomes.

The list of stakeholders that participated in the workshop is summarized in Table 5.4, Attendees at the Interoperable ESE Workshop. Participant

Table 5.3 Interoperable and compatible ESE workshop agenda

	Welcome and introductions	10:00 am
CGrant	(1) Review of workshop goal, objectives, and deliverables	10:05 am
CGrant	(2) Presentation on research study	10:15 am
BVarner	(3) Presentation on prior proposed standardization initiative	10:45 pm
CGrant	(4) Review and preliminary discussion of information presented	11:15 pm
	Networking lunch	12:00 pm
All participants	(5) Discuss/Establish plan for interoperability and proposed standardization	1:00 pm
All participants	(6) Discuss and Clarify plan for tentative proposed radio standard	2:00 pm
All participants	(7) Summarization of prioritized recommended action steps	3:00 pm
	Adjourn	4:00 pm

Table 5.4 Attendees at the interoperable ESE workshop

Attendees interoperable and compatable ESE workshop 20 September 2012 Atlanta GA

First name	Last name	Organization	E-mail
Jason	Allen	Intertek	Jason Allen@lntertek.com
Kamil	Agi	K&A Wireless	kagi@ka-wireless.com
Bob	Athanas	FDNY/SAFE-IR	boba@safe-ir.com
Maxim	Batalin	UCLA Engineering	mbatalini@ita.ucla.edu
Landon	Borders	Bullard	landon borders@bullard.com
Matt	Busa	Motorola Solutions	mattbusa@motorolasGlutions.com
DK	Ezekoye	University of Texas Austin	dezekoye@mail.utexas.edu
Michael	Feely	Boston Fire Dept	Mikef.bfd@cityofboston.gov
Craig	Gestler	Mine Safety ApplianceCompany	Craig.Gestler@MSAsafety.com
Casey	Grant	FPRF	Cgrant@nfpa.org
Paul	Greenberg	NASA	Paul.S.Greenberg@nasa.gov
Beverly	Gul ledge	Scott Safety	bgulledge@tycoint.com
Wayne	Haase	Summit Safety Inc	whaase@summitsafetyinc.com
Zach	Haase	Summit Safety Inc	zhaase@summitsafetyinc.com
Bill	Haskell	NIOSH NPPTL	whaskell@cdc.gov
Simon	Hogg	Draeger	simon.hogg@draeger.com
Jack	Jarboe	Grace Industries	iackjarboepsc@comcast.net
Rich	Katz	Mine Safety Appliance Company	Rich.Katz@MSAsafety.com
Mark	Krizik	Motorola Solutions	mark.krizak@motorolasolutions.com
Bill	Kushner	Motorola Solutions	kushner@motorolasolutions.com
David	Little	ISG [Infrasys	dalittle@mindspring.com
Mark	Mordecai	Globe	mordecai@globefiresuits.com
John	Morris	ISG Infrasys	jmorris@isgfire.com
Craig	Parkulo	Scott Safety	cparkulo@tycoint.com
Kate	Rem ley	NIST	kate.remley@nist.gov
Dan	Rossos	Portland Fire & Rescue (OR)	dan.rossos@portlandoregon.gov
Christina	Spoons	West Dundee Fire Dept (11)	cspoons@msn.com
Joelle	Suits	University of Texas Austin	isuits@utexas.edu
Steven	Townsend	City of Carrollton Fire Dept (TX)	steventownsend@verizon.net
Dave	Trebisacci	NFPA	dtrebisacci@nfpa.org
Bruce	Varner	BHVarner & Associates	Bruce@bhvarner.com
Steve	Weinstein	ISEA (Honeywell Safety Products)	steve.weinstein@honeywell.com
Preston	Wilson	University of Texas Austin	pswilson@mail.utexas.edu
Bill	Young	NIST	wfv@boulder.nist.gov

demographics were intended to include members of the NFPA Technical Committee on Electronic Safety Equipment (ESE) and representatives involved with fireground ESE from the following areas: fire service emergency responders, manufacturers, standards writing representatives, and applicable subject matter experts.

Chapter 6
Recommendations

As a result of the information gathered throughout this project, including the presentations and group discussions from the two workshops, the following recommendations have resulted, in no particular order of priority:

1. **Moving Toward ESE Interoperability**

   1.1. **Supporting an Evolutionary Approach**. Promote concepts that support ESE platforms with individual components that are compatible, integrated, and interoperable. This would be an evolutionary path that recognizes the virtues of a centralized interoperable platform. An example is the combining and maximizing of the efficiency of various features (e.g., power supplies), which would potentially alleviate and mitigate other performance concerns (e.g., insufficient performance due to limited power supplies).

   1.2. **Related Professional Applications**. Identify and consider the lessons learned from professions with parallel ESE applications to structural fire fighting, such as aviation, military, space and underwater diving.

2. **Establishing Central Concepts for ESE Interoperability**

   2.1. **Clarify Definition of ESE**. Clarify the definition of "ESE" to distinguish if it is intended to include or exclude portable, mobile, stationary, and/or field deployable equipment.

   2.2. **Define ESE Interoperability**. Define "ESE Interoperability" to distinguish it from fireground interoperability and wireless communication interoperability. A possible definition is: "ESE Interoperability—the ability of ESE to operate in synergy in the execution of assigned tasks."

   2.3. **ESE Categories**. Consider categorization of emergency responder ESE, such as:

   (a) Communications
   (b) Environmental monitoring
   (c) Physiological monitoring
   (d) Sensory support
   (e) Tracking/location

C. C. Grant, *Interoperable Electronic Safety Equipment*,
SpringerBriefs in Fire, DOI: 10.1007/978-1-4614-8277-2_6,
© Fire Protection Research Foundation 2012

2.4. **Responder Knowledge Base**. Continue to recognize, utilize, and support the Responder Knowledge Base as a mechanism for tracking available ESE.

2.5. **Interoperability Performance Characteristics**. Consider the key interoperability performance characteristics for fire service ESE as electrical oriented and non-electrical oriented. Examples of electrical oriented performance characteristics include:

(a) Inter-component communication
(b) Centralized power supply and distribution
(c) Non-interference

Examples of non-electrical oriented performance characteristics include:

(a) Form, fit and function
(b) Ergonomics
(c) User interface
(d) Donning and doffing

2.6. **Component Attributes**. Consider the primary ESE component attributes, which are:

(a) Operability
(b) Maintainability
(c) Durability
(d) Availability
(e) Stability
(f) Reliability

3. **ESE Interoperability Standardization**

3.1. **Standardize Interoperability Concepts**. Document interoperability concepts in consensus-developed codes and standards documents. Use these documented requirements and/or guidelines to provide an appropriate baseline to address the overall topic of interoperability.

3.2. **Define the Fire Service Landscape**. Better define the requirements for fire service ESE by clarifying fireground environments and fire fighter needs, with specific attention to how ESE will be used in different situations. Transpose this information into the requirements or guidelines in standardization documents.

3.3. **Consistency of Requirements**. Consistency of performance requirements across all emergency responder ESE is a sensible goal, and consideration of logical differences in performance requirements should be based on substantive technical rationale. Action items that should be considered include:

3.3.1. Revisit NFPA requirements for performance requirements for all ESE, using an approach similar to the recent analysis provided for PASS by the Intrinsic Safety Task Group for the NFPA ESE Technical Committee.

3.3.2. Consider this effort through the PPE Correlating Committee since it affects multiple Technical Committees under their direction.

3.4. **Periodic Re-Evaluation**. The performance characteristics for different ESE should be re-evaluated on a periodic basis, since the technological landscape is continually changing and subject to ongoing advancements that impact the respective requirements.

4. **Intrinsic Safety of ESE**

4.1. **Periodic Re-Evaluation**. The need for intrinsic safety requirements for different ESE should be re-evaluated on a periodic basis, since the technological landscape is continually changing and subject to on-going advancements that impact the respective requirements.

4.2. **Interoperability**. Consideration should be given to promote concepts of interoperability, since a centralized interoperable platform that combines and maximizes the efficiency of multiple power supplies would potentially also alleviate other concerns (e.g., the current trade-off for reduced power supplies to comply with more rigorous Division 1 requirements).

4.3. **Consistency of Requirements**. Consistency of intrinsic safety requirements across all emergency responder ESE is a sensible goal that should be founded on the inherent technological differences of ESE that justify different intrinsic safety requirements. Action items that should be considered include:

4.3.1. NFPA requirements for intrinsic safety should be revisited and considered for all ESE, using an approach similar to the recent analysis provided for PASS by the Intrinsic Safety Task Group for the NFPA ESE Technical Committee.

4.3.2. This effort should be considered by the PPE Correlating Committee since it affects multiple Technical Committees under their direction.

4.4. **Defining the Fire Service Landscape**. Better define the requirements for intrinsically safe ESE by clarifying fireground environments and fire fighter needs. Examples of factors that should be considered are:

4.4.1. Division 1 and Division 2 levels of safety are presently based on normal or abnormal probability of occurrence, but this is meant for the electrical equipment installed at a fixed location rather than portable equipment moving in and out of hazardous locations.

4.4.2. A likely scenario for fire fighters is a hazardous environment where it's not expected, such as a gas leak, and in those cases the equipment on-site (without intrinsic safety protection) would arguably introduce a more realistic danger than the fire fighters own equipment.

4.4.3. Since intrinsic safety is generally applicable across all applications, clarify the unique features of a fireground with other applications (e.g., petrochemical).

4.5. **Ongoing Dialogue**. Further dialogue should be facilitated among the intrinsic safety product standards developers (i.e., FM, TIA, UL, etc.), manufacturers, and the user community, to clarify additional details needed for the proper references between the enabling standards (i.e., NFPA) and the product standards.

Appendix A
Presentations from Intrinsic Safety Workshop

Annex A provides a compilation of the presentations provided at the one-day workshop held in Northbrook, Illinois, on Wednesday, 19 September 2012. The purpose of these presentations was to address intrinsic safety of portable electronic safety equipment used by emergency responders. A summary of the presentations in this annex is provided in Table A.1, Summary of Intrinsic Safety Workshop Presentations. The PowerPoint presentations are included in the following pages in the order they were given at the workshop.

Table A.1 Summary of intrinsic safety workshop presentations

Figures	Description	No. of pages
Figures A.1, A.2, A.3 and A.4	Presentation by Casey Grant, Fire Protection Research Foundation	4
Figure A.5	Presentation by Paul Kelly, Underwriters Laboratories	1
Figures A.6 and A.7	Presentation by Robert Martell, Factory Mutual Approvals	2

C. C. Grant, *Interoperable Electronic Safety Equipment*,
SpringerBriefs in Fire, DOI: 10.1007/978-1-4614-8277-2,
© Fire Protection Research Foundation 2012

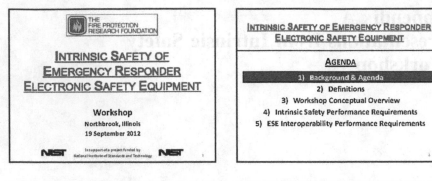

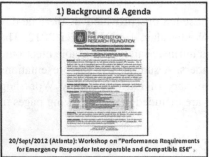

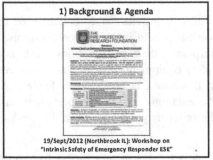

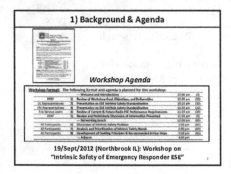

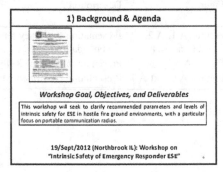

Fig. A.1 FPRF presentation (sheet 1 of 4)

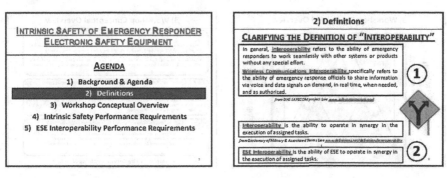

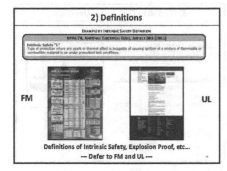

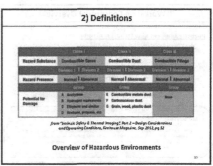

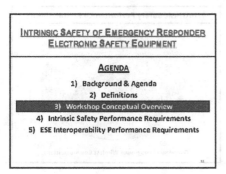

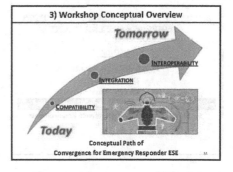

Fig. A.2 FPRF presentation (sheet 2 of 4)

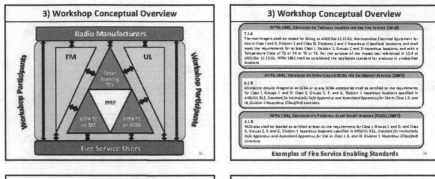

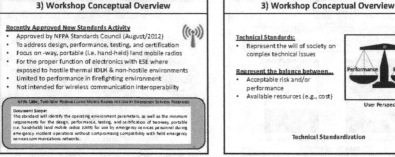

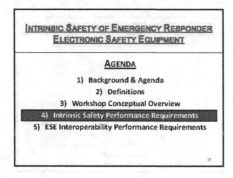

Fig. A.3 FPRF presentation (sheet 3 of 4)

Overview of Fireground Intrinsic Safety

Operational Hazards to/from ESE

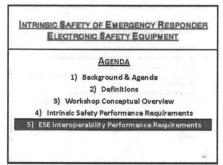

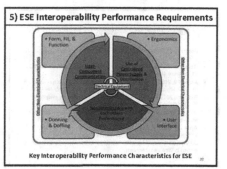

Key Interoperability Performance Characteristics for ESE

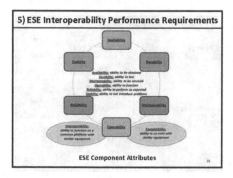

ESE Component Attributes

Fig. A.4 FPRF presentation (sheet 4 of 4)

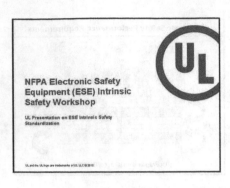

NFPA Electronic Safety Equipment (ESE) Intrinsic Safety Workshop

UL Presentation on ESE Intrinsic Safety Standardization

UL and the UL logo are trademarks of UL LLC © 2013

NFPA ESE-related standards

- ANSI/NFPA 1801:2013 on Thermal Imagers for the Fire Service
- ANSI/NFPA 1852:2008 and 1981:2007 on Self-Contained Breathing Apparatus (SCBA)
- ANSI/NFPA 1982:2007 on Personal Alert Safety Systems (PASS)
- Draft NFPA 1800 on ESE for Emergency Services
- New NFPA effort on Land Mobile Radios (LMRs)
- New NFPA effort on GPS Locator Trackers

Note: ANSI/NFPA 1981:2007 and 1982:2007 both include references to ANSI/UL 913, 7th edition, "Intrinsically Safe Apparatus and Associated Apparatus for Use in Class I, II, and III, Division 1, Hazardous (Classified) Locations".

Recent ANSI/UL 913 History

- **February 21, 1997:** ANSI/UL 913, 5th edition, published and effective.
- **Between 1997 & 2002:** Industry begins discussions on revising UL 913 to more closely align with IEC 60079-11.
- **August 8, 2002:** ANSI/UL 913, 6th edition published with revisions to more closely align with IEC 60079-11 (*not for any safety reasons*).
- **Between 2002 & 2006:** Industry begins discussion on revising UL 913 to fully harmonize with IEC 60079-11.
- **July 31, 2006:** ANSI/UL 913, 7th edition, published with revisions to fully harmonize with IEC 60079-11 (*not for any safety reasons*).
- **April 25, 2008:** ANSI/UL 913, 6th edition withdrawn based on 7th edition publication prior to 6th edition becoming effective.
- **July 31, 2016:** ANSI/UL 913, 7th edition, to become effective, with 5th edition to then be withdrawn.

Key Differences From 5th to 7th Editions

- Dust-tight enclosures for Group E applications.
- Electrostatic requirements for non-metallic enclosures.
- Drop testing at low ambient temperatures.
- Revised spark ignition test factor from 1.22 to 1.5 on voltage or current.
- Analysis to determine maximum temperatures.
- De-rating under fault conditions without testing option.
- Encapsulation and ratings for fuses.
- Electrolyte leakage & temperature testing of batteries.
- Compliance with UL 2054 abnormal conditions for battery packs.

Battery-Powered, Portable Land Mobile Radios

- Telecommunications Industry Association (TIA) published an LMR standard for HazLoc applications.
- TIA 4950, "Requirements For Battery-powered, Portable Land Mobile Radio Applications In Class I, II, And III, Division 1, Hazardous (Classified) Locations".
- ANSI recognition is pending for TIA 4950.
- Requirements in TIA 4950 are based on ANSI/UL 913, 5th edition.
- UL is actively in support of this effort by TIA.
- Scope of accreditation for UL is being revised to include TIA 4950.
- Certification efforts based on TIA 4950 are underway by UL.

THANK YOU.

Fig.A.5 UL presentation (sheet 1 of 1)

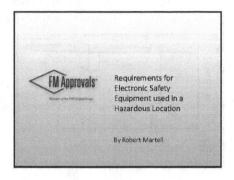

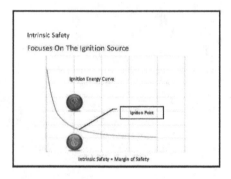

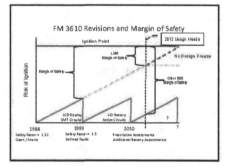

Fig. A.6 FM presentation (sheet 1 of 2)

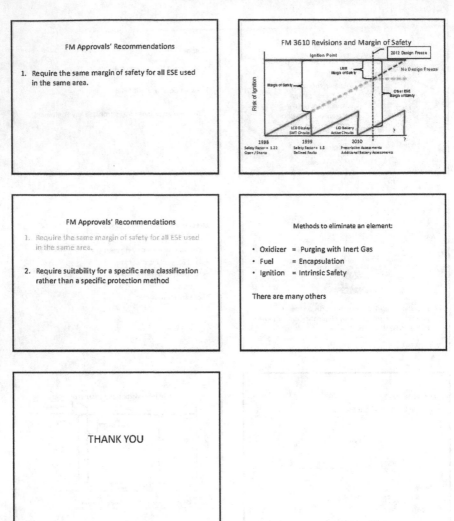

Fig. A.7 FM presentation (sheet 2 of 2)

Appendix B
Presentations from Interoperability Workshop

Annex B provides a compilation of the presentations provided at the one-day workshop held in Atlanta, Georgia on Thursday, 20 September 2012, whose purpose was to address interoperability of portable electronic safety equipment used by emergency responders. A summary of the presentations in this annex is provided in Table B.1, Summary of Interoperability Workshop Presentations. The PowerPoint presentations are included in the following pages in the order they were given at the workshop.

Table B.1 Summary of interoperability workshop presentations

Figures	Description	No. of pages
Figures B.1, B.2, B.3, B.4, B.5, B.6 B.7 and B.8	Presentation by Casey Grant, Fire Protection Research Foundation	8
Figures B.9, B.10, B.11 and B.12	Presentation by Bill Kushner, Motorola Solutions	4
Figures B.13, B.14, B.15, B.16, B.17 and B.18	Presentation by Preston Wilson, University of Texas—Austin	6
Figures B.19, B.20, B.21 and B.22	Presentation by Maxim Batalin, UCLA Engineering	4
Figures B.23, B.24 and B.25	Presentation by Paul Greenberg, NASA	3

C. C. Grant, *Interoperable Electronic Safety Equipment*,
SpringerBriefs in Fire, DOI: 10.1007/978-1-4614-8277-2,
© Fire Protection Research Foundation 2012

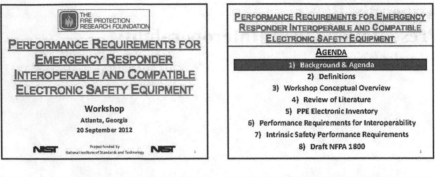

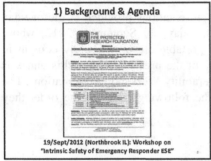

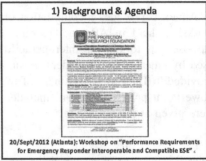

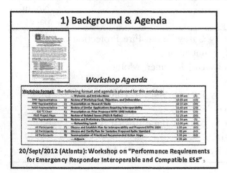

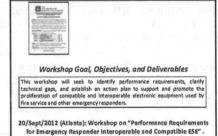

Fig. B.1 FPRF presentation (sheet 1 of 8)

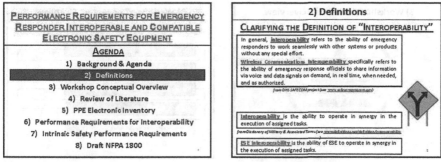

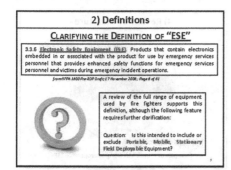

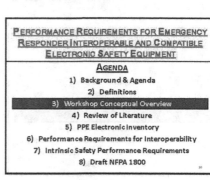

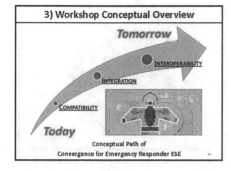

Fig. B.2 FPRF presentation (sheet 2 of 8)

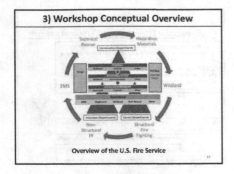

Overview of the U.S. Fire Service

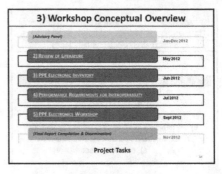

Project Tasks

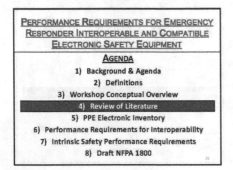

Examples of Professions Using Interoperable ESE

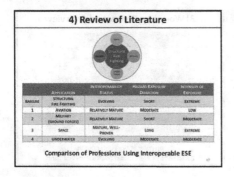

Comparison of Professions Using Interoperable ESE

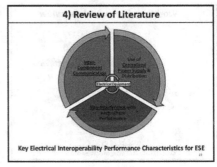

Key Electrical Interoperability Performance Characteristics for ESE

Fig. B.3 FPRF presentation (sheet 3 of 8)

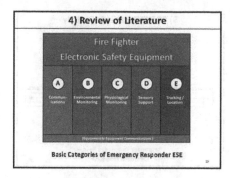

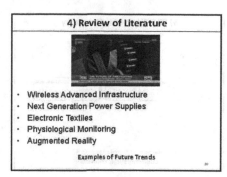

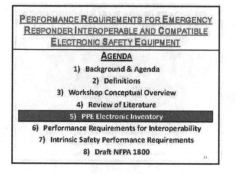

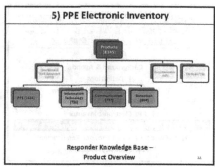

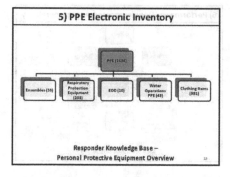

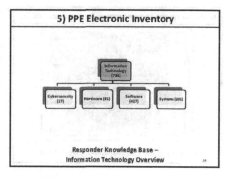

Fig. B.4 FPRF presentation (sheet 4 of 8)

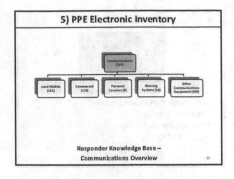

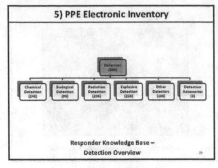

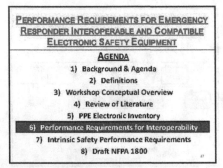

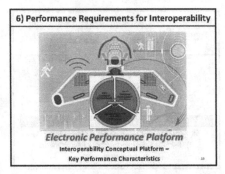

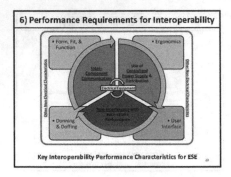

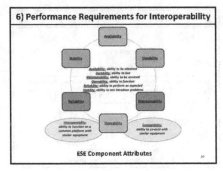

Fig. B.5 FPRF presentation (sheet 5 of 8)

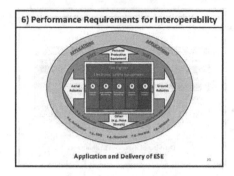

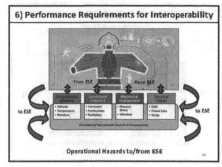

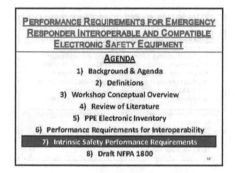

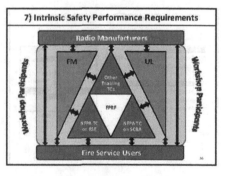

Fig. B.6 FPRF presentation (sheet 6 of 8)

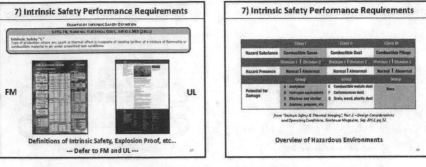

7) Intrinsic Safety Performance Requirements

Definitions of Intrinsic Safety, Explosion Proof, etc...
--- Defer to FM and UL ---

7) Intrinsic Safety Performance Requirements

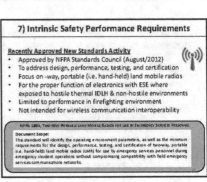

Overview of Hazardous Environments

7) Intrinsic Safety Performance Requirements

Examples of Fire Service Enabling Standards

7) Intrinsic Safety Performance Requirements

Recently Approved New Standards Activity
- Approved by NFPA Standards Council (August/2012)
- To address design, performance, testing, and certification
- Focus on -way, portable (i.e. hand-held) land mobile radios
- For the proper function of electronics with ESE where exposed to hostile thermal IDLH & non-hostile environments
- Limited to performance in firefighting environment
- Not intended for wireless communication interoperability

NFPA 1800X, TWO-WAY PORTABLE LAND MOBILE RADIOS FOR USE BY EMERGENCY SERVICE PERSONNEL

Document Scope:
This standard will identify the operating environment parameters, as well as the minimum requirements for the design, performance, testing, and certification of two-way, portable (i.e. hand-held) land mobile radios (LMR) for use by emergency services personnel during emergency incident operations without compromising compatibility with field emergency services communications networks.

7) Intrinsic Safety Performance Requirements

Technical Standards:
- Represent the will of society on complex technical issues

Represent the balance between...
- Acceptable risk and/or performance
- Available resources (e.g., cost)

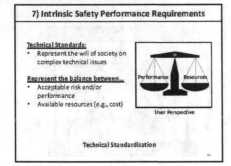

User Perspective

Technical Standardization

PERFORMANCE REQUIREMENTS FOR EMERGENCY RESPONDER INTEROPERABLE AND COMPATIBLE ELECTRONIC SAFETY EQUIPMENT

AGENDA

1) Background & Agenda
2) Definitions
3) Workshop Conceptual Overview
4) Review of Literature
5) PPE Electronic Inventory
6) Performance Requirements for Interoperability
7) Intrinsic Safety Performance Requirements
8) Draft NFPA 1800

Fig. B.7 FPRF presentation (sheet 7 of 8)

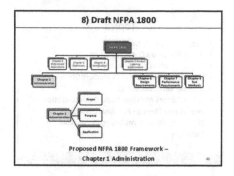

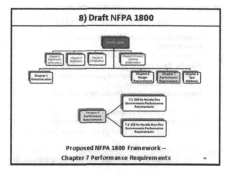

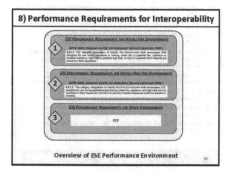

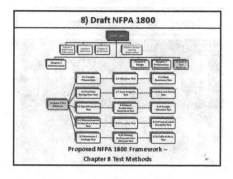

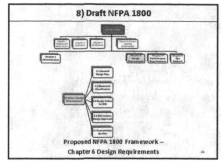

Fig. B.8 FPRF presentation (sheet 8 of 8)

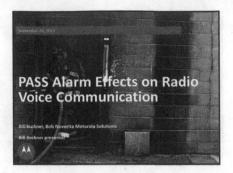

Topics

- Fireground Safety Goals
- Alarm Effects on Communication
- Speech Production Refresher
- Speech and PASS Alarm Characteristics
- How Do Vocoders Work?
- Alarm Factors Affecting Radio Communication
- PASS Alarm Through P25 Radio Examples
- Suggested PASS Alarm Design Constraints
- Summary

Fireground Safety Goals

- **Prevent physical injury due to the work environment**
  - Turnout gear, SCBA, appropriate tools, etc.
- **Alert for dangerous user/work conditions**
  - PASS, low-air alarms, heat and gas sensors, physical condition monitors, etc.
- **Insure local/fire scene situational awareness**
  - Personnel accountability
  - Knowledge of scene geography
  - Reliable communication with command and squad
- All components are part of the safety "system" and must work together

Alarm Noise Effects on Communication

- Noise diminishes speech intelligibility by masking speech (producing low relative SNR in frequency and time).
- Effective alarm signals draw attention by stimulating targeted human hearing sensitivities and using sound levels above external noise masking levels.
- Acoustic alarms and speech share the same audio band. Alarm "noise" can mask important speech signal components.
- In addition, noise can interfere with the radio speech encoding process in both a linear and non-linear manner.
- Certain audio characteristics that make the best alarm signals can worsen radio speech encoding and overall intelligibility.
- New alarm signal designs should take into account radio system vulnerabilities.

Speech Production Refresher

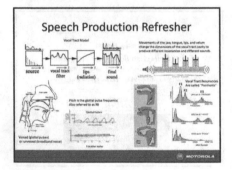

Speech Signal and Spectrogram

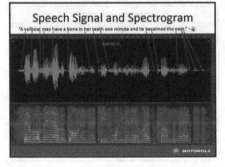

Fig. B.9 Motorola presentation (sheet 1 of 4)

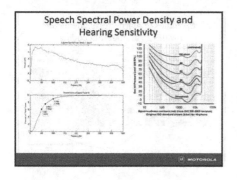

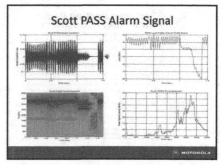

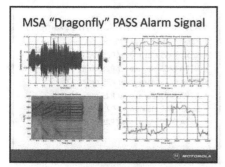

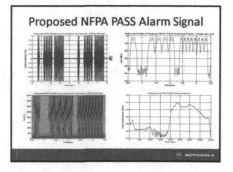

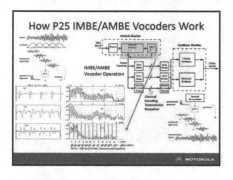

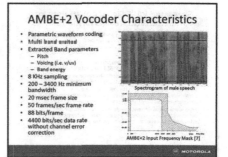

Fig. B.10 Motorola presentation (sheet 2 of 4)

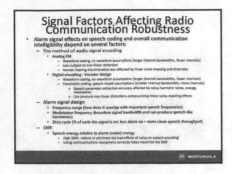

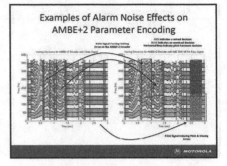

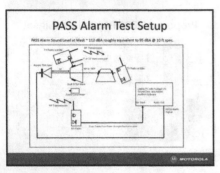

PASS Alarm P25 Radio TX/RX Examples

- Clean speech over P25 radio, 1" —
- Scott PASS alarm over P25 radio, 1" —
- NFPA PASS alarm over P25 radio, 1" -
- Scott PASS alarm over P25 radio, 12" —
- NFPA PASS alarm over P25 radio, 12" —
- Scott PASS alarm over P25 radio, RSM —
- NFPA PASS alarm over P25 radio, RSM —

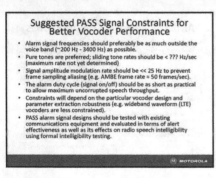

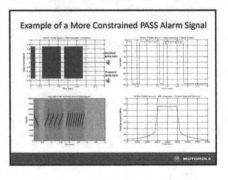

Fig. B.11 Motorola presentation (sheet 3 of 4)

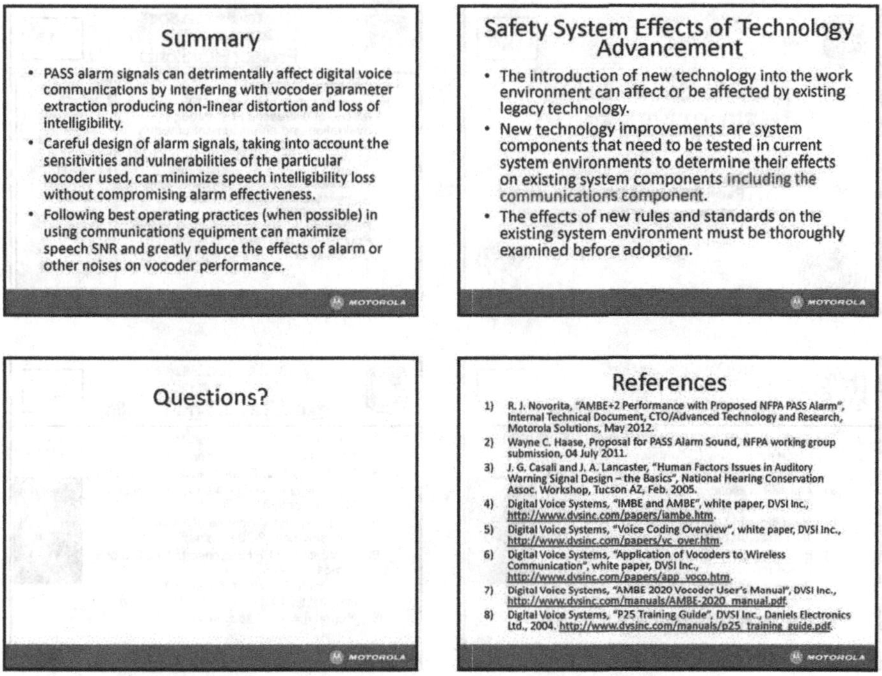

Summary

- PASS alarm signals can detrimentally affect digital voice communications by interfering with vocoder parameter extraction producing non-linear distortion and loss of intelligibility.
- Careful design of alarm signals, taking into account the sensitivities and vulnerabilities of the particular vocoder used, can minimize speech intelligibility loss without compromising alarm effectiveness.
- Following best operating practices (when possible) in using communications equipment can maximize speech SNR and greatly reduce the effects of alarm or other noises on vocoder performance.

Safety System Effects of Technology Advancement

- The introduction of new technology into the work environment can affect or be affected by existing legacy technology.
- New technology improvements are system components that need to be tested in current system environments to determine their effects on existing system components including the communications component.
- The effects of new rules and standards on the existing system environment must be thoroughly examined before adoption.

Questions?

References

1) R. J. Novorita, "AMBE+2 Performance with Proposed NFPA PASS Alarm", Internal Technical Document, CTO/Advanced Technology and Research, Motorola Solutions, May 2012.
2) Wayne C. Haase, Proposal for PASS Alarm Sound, NFPA working group submission, 04 July 2011.
3) J. G. Casali and J. A. Lancaster, "Human Factors Issues in Auditory Warning Signal Design – the Basics", National Hearing Conservation Assoc. Workshop, Tucson AZ, Feb. 2005.
4) Digital Voice Systems, "IMBE and AMBE", white paper, DVSI Inc., http://www.dvsinc.com/papers/iambe.htm.
5) Digital Voice Systems, "Voice Coding Overview", white paper, DVSI Inc., http://www.dvsinc.com/papers/vc_over.htm.
6) Digital Voice Systems, "Application of Vocoders to Wireless Communication", white paper, DVSI Inc., http://www.dvsinc.com/papers/app_voco.htm.
7) Digital Voice Systems, "AMBE 2020 Vocoder User's Manual", DVSI Inc., http://www.dvsinc.com/manuals/AMBE-2020_manual.pdf.
8) Digital Voice Systems, "P25 Training Guide", DVSI Inc., Daniels Electronics Ltd., 2004. http://www.dvsinc.com/manuals/p25_training_guide.pdf.

Fig. B.12 Motorola presentation (sheet 4 of 4)

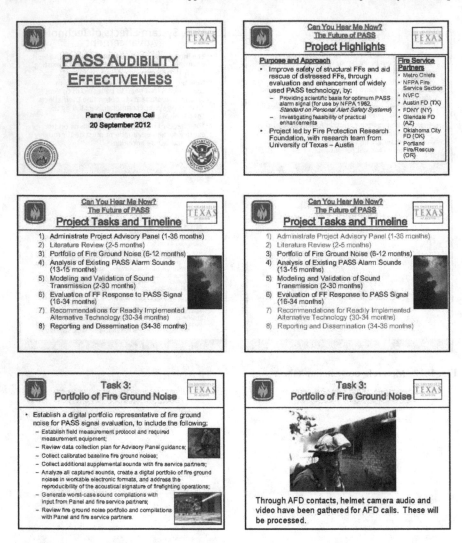

Fig. B.13 University of Texas—Austin presentation (sheet 1 of 6)

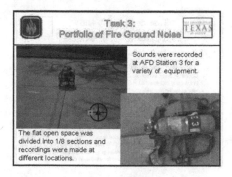

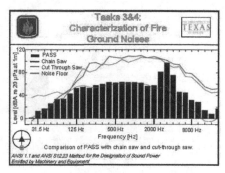

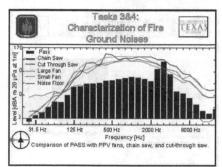

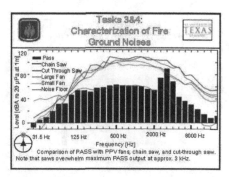

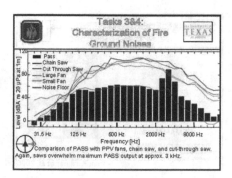

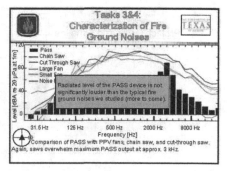

Fig. B.14 University of Texas—Austin presentation (sheet 2 of 6)

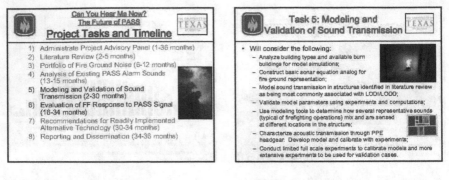

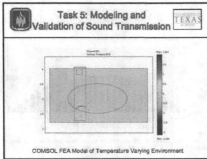

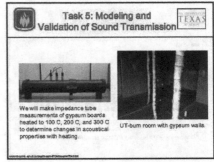

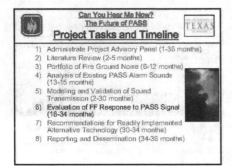

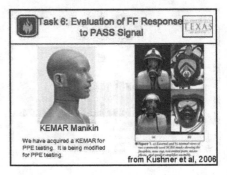

Fig. B.15 University of Texas—Austin presentation (sheet 3 of 6)

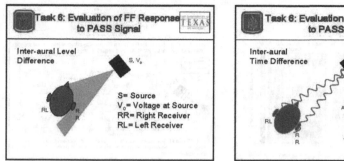
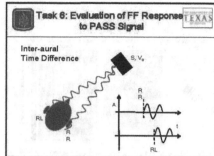

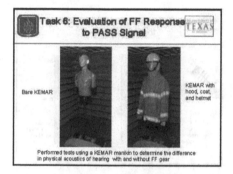
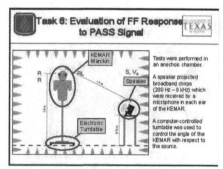

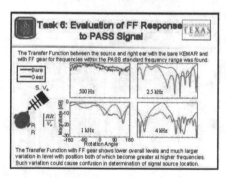
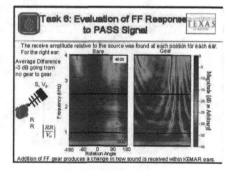

Fig. B.16 University of Texas—Austin presentation (sheet 4 of 6)

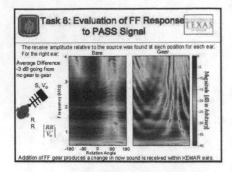

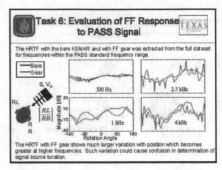

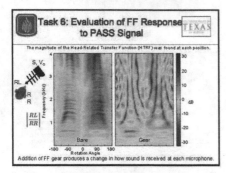

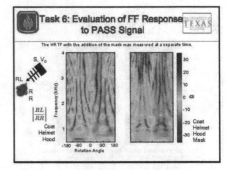

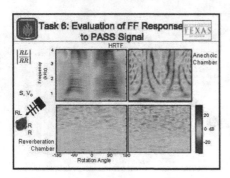

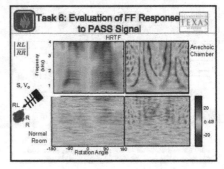

Fig. B.17 University of Texas—Austin presentation (sheet 5 of 6)

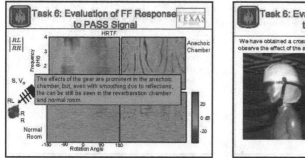

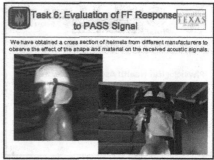

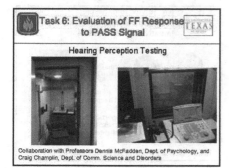

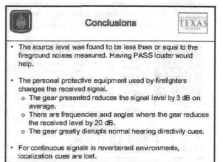

Fig. B.18 University of Texas—Austin presentation (sheet 6 of 6)

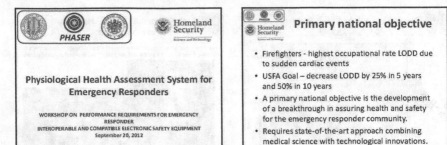

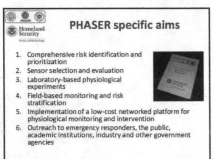

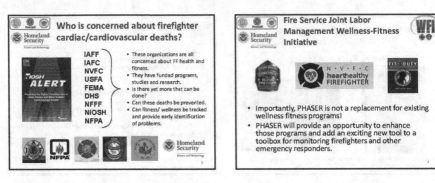

Fig. B.19 UCLA Engineering presentation (sheet 1 of 4)

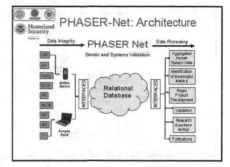

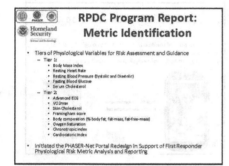

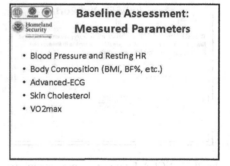

Fig. B.20 UCLA Engineering presentation (sheet 2 of 4)

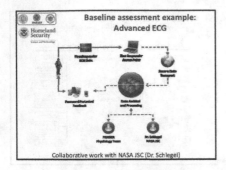

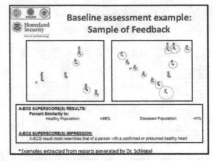

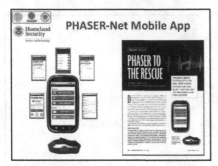

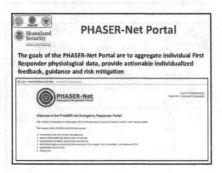

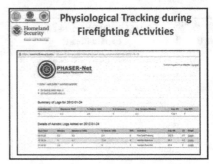

Fig. B.21 UCLA Engineering presentation (sheet 3 of 4)

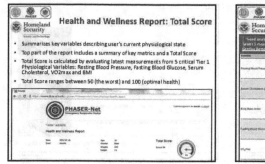

Health and Wellness Report: Total Score

Homeland Security

- Summarizes key variables describing user's current physiological state
- Top part of the report includes a summary of key metrics and a Total Score
- Total Score is calculated by evaluating latest measurements from 5 critical Tier 1 Physiological Variables: Resting Blood Pressure, Fasting Blood Glucose, Serum Cholesterol, VO2max and BMI
- Total Score ranges between 50 (the worst) and 100 (optimal health)

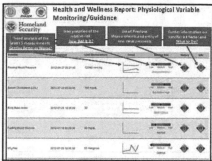

Health and Wellness Report: Physiological Variable Monitoring/Guidance

Homeland Security

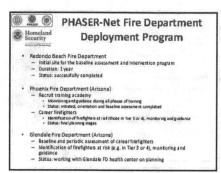

PHASER-Net Fire Department Deployment Program

Homeland Security

- Redondo Beach Fire Department
  - Initial site for the baseline assessment and intervention program
  - Duration: 1 year
  - Status: successfully completed
- Phoenix Fire Department (Arizona)
  - Recruit training academy
    - Monitoring and guidance during all phases of training
    - Status: initiated, orientation and baseline assessment completed
  - Career firefighters
    - Identification of firefighters at risk (those in Tier 3 or 4), monitoring and guidance
    - Status: final planning stages
- Glendale Fire Department (Arizona)
  - Baseline and periodic assessment of career firefighters
  - Identification of firefighters at risk (e.g. in Tier 3 or 4), monitoring and guidance
  - Status: working with Glendale FD health center on planning

PHASER-Net Platform Support for Other Promising Future Capabilities

Homeland Security

- EMS rehabilitation dashboard: To provide decision support for EMS personnel during rehabilitation and beyond
- Support of the optimized Incident Commander monitoring and control interface
  - Integration of location and physiological information
  - Compact presentation of data enabling decision support
  - Task assignment tool based on physiological state
- On-scene monitoring and evaluation of wearable ECG
  - Application of automated advanced ECG algorithms
  - Digital filtering to remove motion and other artifacts
- Direct risk diagnostics based on lab-on-a-chip sensor platforms
  - Direct measurements of changes in cellular properties
  - Direct accurate and sensitive measures of known biomarkers
  - Current prototypes enable this technology to be low-cost, expedient and accurate

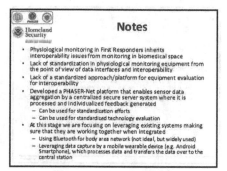

Notes

Homeland Security

- Physiological monitoring in First Responders inherits interoperability issues from monitoring in biomedical space
- Lack of standardization in physiological monitoring equipment from the point of view of data interfaces and interoperability
- Lack of a standardized approach/platform for equipment evaluation for interoperability
- Developed a PHASER-Net platform that enables sensor data aggregation by a centralized secure server system where it is processed and individualized feedback generated
  - Can be used for standardization efforts
  - Can be used for standardized technology evaluation
- At this stage we are focusing on leveraging existing systems making sure that they are working together when integrated
  - Using Bluetooth for body area network (not ideal, but widely used)
  - Leveraging data capture by a mobile wearable device (e.g. Android Smartphone), which processes data and transfers the data over to the central station

Fig. B.22 UCLA Engineering presentation (sheet 4 of 4)

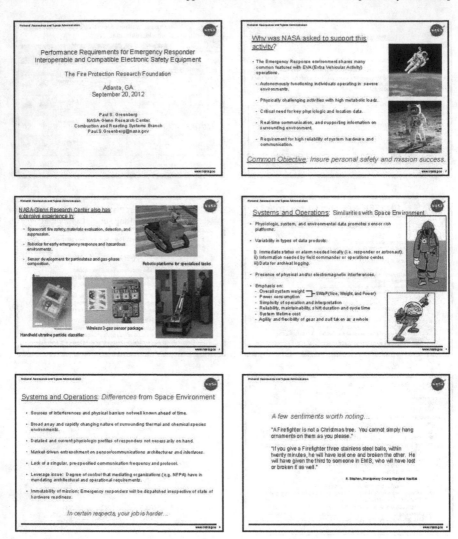

Fig. B.23 NASA presentation (sheet 1 of 3)

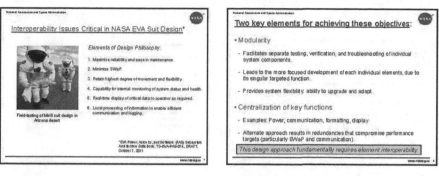

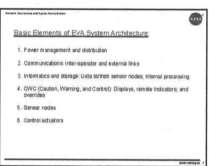

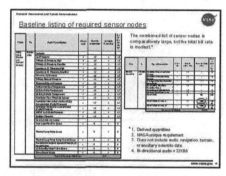

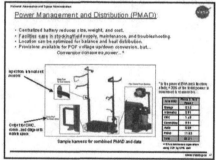

Fig. B.24 NASA presentation (sheet 2 of 3)

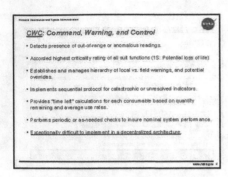

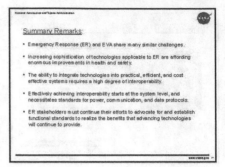

Fig. B.25 NASA presentation (sheet 3 of 3)

Bibliography

The following is a summary of the reference sources cited throughout this report and other readily available literature identified throughout this study addressing the subject matter of interest.

3-Up battery charging portable power system for military electronic equipment. Fuel Cell Ind. Rep. **11**(6), 1 (2010)

H.S. Abdelsalam, in *Energy Efficient Workforce Selection in Special-Purpose Wireless Sensor Networks*, Proceedings of IEEE INFOCOM Workshops, 2008

F. Amon, N. Bryner, A. Hamins, in *Thermal Imaging Research Needs for First Responders: Workshop Proceedings*, NIST Special Publication 1040, National Institute of Standards and Technology, June 2005

F. Amon, A. Hamins, N. Bryner, J. Rowe, Meaningful performance evaluation conditions for fire service thermal imaging cameras. Fire Saf. J. **43**(8), 541–550 (2008)

F.K. Amon, D. Leber, N. Paulter, in *Objective Evaluation of Imager Performance*, Fifth International Conference on Sensing Technology, Dec 2011, pp. 47–52

M Angermann, M. Khider, P. Robertson, in *Towards Operational Systems for Continuous Navigation of Rescue Teams*, 2008 IEEE/ION Position Location and Navigation Symposium, May 2008, pp. 153–158

D. Annapurna, D. Shreyas Bhagavath, V. Gnanaskandan, K.B. Raja, K.R. Venugopal, L.M. Patnaik, Performance comparison of AODV, AOMDV and DSDV for fire fighters application. Commun. Comput. Inf. Sci. **250**(Part 1) 363–367 (2011)

ANSI/ISA-S82.03, *Safety Standard for Electrical and Electronic Test, Measuring, Controlling and Related Equipment* (1988)

Avionics fly to fire fighters rescue. Mach. Des. **65**(20), 16 (1993)

C. Baber, D.J. Haniff, R. Buckley, Wearable information appliances for the emergency services: HotHelmet. Lect. Notes Comput. Sci. **1707**, 314–316 (1999)

D. Barr, T. Reilly, W. Gregson, The impact of different cooling modalities on the physiological responses in firefighters during strenuous work performed in high environmental temperatures. Eur. J. Appl. Physiol. **111**(6), 959–967 (2011)

A. Bonfiglio et al., Managing catastrophic events by wearable mobile systems. Lect. Notes Comput. Sci. **4458**, 95–105 (2007)

H.R. Booher, *Handbook of Human Systems Integration* (Wiley, Hoboken, 2003)

N. Bretschneider, S. Brattke, R. Karlheinz, in *Head Mounted Display for Fire Fighters*, 3rd International Forum on Applied Wearable Computing, Mar 2006, pp. 1–15

R.A. Bryant, K.M Butler, R.L. Vettori, P.S. in *Greenberg, Real-Time Particulate Monitoring— Detecting Respiratory Threats for First Responders: Workshop Proceedings*, NIST Special Publication 1051, National Institute of standards and Technology, Gaithersburg MD, Dec 2007

C. C. Grant, *Interoperable Electronic Safety Equipment*,
SpringerBriefs in Fire, DOI: 10.1007/978-1-4614-8277-2,
© Fire Protection Research Foundation 2012

N. Bryner, D. Madrzykowski, D. Stroup, in *Performance of Thermal Exposure Sensors in Personal Alert Safety System (PASS) Devices*, NISTIR 7295, National Institute of Standards and Technology, Sept 2005

S. Bush, Mini fuel cell lasts 72 hrs. *Electronics Weekly*, **2210**, 7 (2005)

M. Buyan, P. Bruhwiler, A. Azens, G. Gustavsson, R. Karmhag, C. Granqvist, Facial warming and tinted helmet visors. Int. J. Ind. Ergon. **36**(1), 11–16 (2006)

J. Chou, *Hazardous Gas Monitors: A Practical Guide to Selection, Operation and Applications* (McGraw-Hill Book Company, NY, 2000)

Committee on Homeland and National Security, Subcommittee on Standards, *A National Strategy for CBRNE Standards*, National Science and Technology Council, May 2011

Committee on Using Information Technology to Enhance Disaster Management, *Summary of a Workshop on Using Information Technology to Enhance Disaster Management* (National Academies Press, 2005)

D. Copeland, WPI devices help locate firefighters. Boston Globe (2009)

W.D. Davis, M.K. Donnelly, M.J. Selepak, N. Building, in *Testing of Portable Radios in a Fire Fighting Environment*, NISTIR 1477, National Institute of Standards and Technology, Building and Fire Research Laboratory, 2006

D. Dawson, Fast fielding for soldier systems. Soldiers **59**(3), 42 (2004)

J.R. Delaney, B. John, United States Patent Application 20090072966, Enhanced Firefighter Door Chock, 19 March 2009, website: http://appft.uspto.gov/netacgi/nph-Parser?Sect1=PTO1 &Sect2=HITOFF&p=1&u=%2Fnetahtml%2FPTO%2Fsrchnum.html&r=1&f=G&l=50&d= PG01&s1=12208339

R.I. Desourdis, *Achieving Interoperability in Critical IT and Communication Systems* (Artech House Inc, Boston, 2009)

S. Dogra, S. Manna, A. Banik, S. Maiti, S.K. Sarkar, in *A Novel Approach for RFID Based Fire Protection*, International Conference on Emerging Trends in Electronic and Photonic Devices & Systems, 2009, pp. 198–201

M.K. Donnelly, W.D. Davis, J.R. Lawson, M.J. Selepak, *Thermal Environment for Electronic Equipment Used by First Responders, NIST Technical Note 1474* (National Institute of standards and Technology, Gaithersburg MD, 2006)

T. Donnelly, Building collapse rescue operations: technical search capabilities. Fire Eng. **163**(10), 22–26 (2010)

J.R. Duckworth, Tracking lost firefighters: firefighter search & rescue systems demonstrated at WPI workshop. Firehouse Mag. **35**(10), 94 (2010)

F. Durso, Sending and receiving. NFPA J. **105**(7), 11 (2011)

R.F. Fahy, P.R. LeBlanc, J.L. Molis, *Firefighter Fatalities in the United States—2008* (NFPA, Quincy, 2009)

G.R. Faulhaber, Solving the interoperability problem: are we on the same channel? An essay on the problems and prospects for public safety radio. Fed. Commun. Law J. **59**(3), 493 (2007)

A.A. Fatah et al., *An Introduction to Biological Agent Detection Equipment for Emergency First Responders, NIJ Guide 101–00* (National Institute of Justice, Washington DC, 2001)

A.A. Fatah, *Guide for the Selection of Communication Equipment for Emergency First Responders, NIJ Guide 104–00* (National Institute of Justice, Washington DC, Volume I, Feb, 2002)

D.Y. Fei, X. Zhao, C. Boanca, E. Hughes, O. Bai, R. Merrell, A. Rafiq, A biomedical sensor system for real-time monitoring of astronauts' physiological parameters during extra-vehicular activities. Comput. Biol. Med. **40**(7), 635–642 (2010)

C. Fischer, H. Gellersen, Location and navigation support for emergency responders: a survey. IEEE Pervasive Comput. **9**(1), 38–47 (2010)

S. Foster, GPS system is lifeline to firefighters. Comput. Canada **30**(4), 16 (2004)

FM 3610, *Approval Standard for Intrinsically Safe Apparatus and Associated Apparatus for Use in Class I, II, and III, Divisions 1, Hazardous (Classified) Locations*, Jan 2010

P. Frazier, R. Hooper, B. Orgen, in *Current Status, Knowledge Gaps, and Research Needs Pertaining to Firefighter Radio Communication Systems*, NIOSH, Morgantown, WV, and TriData Corporation, Arlington, VA, Sept 2003

Fuel Cell Today Industry Review 2011, Johnson Matthey PLC, July 2011, website: http://www.fuelcelltoday.com/media/1351623/industry_review_2011.pdf, cited: 8 Feb 2011, cited: 6 Feb 2012

FireRescue1Staff, Futuristic New Helmet Helps Firefighters See Through Smoke, FireRescue1, website: www.firerescue1.com/fire-products/personal-protective-equipment-ppe/helmets/articles/1315543-Futuristic-new-helmet-helps-firefighters-see-through-smoke/, cited: 31 July 2012

H. Goldstein, Radio contact in high-rises can quit on firefighters. IEEE Spectr. **39**(4), 24–27 (2002)

C.C. Grant, *Respiratory Exposure Study for Fire Fighters and Other Emergency Responders* (Fire Protection Research Foundation, Quincy, 2007)

C.C. Grant, The future of fire: research renaissance? NFPA J. Quincy, MA, Mar/Apr 2009

J.R. Hall, *The Total Cost of Fire in the United States* (NFPA, Quincy, MA, 2011)

C. Hartung, R. Han, C. Seielstad, S. Holbrook, FireWxNet: A multi-tiered portable wireless system for monitoring weather conditions in wildland fire environments. Paper presented at Proceedings of the 4th International Conference on Mobile Systems, Applications and Services, 2006

C. Hawley, in *Hazardous Materials Air Monitoring & Detection Devices*, 2nd edn. (Thomas Delmar Learning, 2007)

M. Hopmeier, H. Christen, M. *Malone, Development of Human Factors Engineering Requirements for Fire Fighting Protective Equipment* (Army Research Development and Engineering Command, Natick Soldier Center, Unconventional Concepts Inc., 2005)

T.B. House, R.E. Strunk, Army soldier enhancement program. Army Sustainment **43**(1), 20 (2011)

Improving Disaster Management: The Role of IT in Mitigation, Preparedness, Response, and Recovery (National Academies Press, 2007)

Intrinsic Safety & Thermal Imaging: Part 2—Design Considerations and Operating Conditions. Firehouse Mag. 52 (Sep 2012)

S. Jayaraman, P. Kiekens, A.M. Grancaric, in *Intelligent Textiles for Personal Protection and Safety*, NATO Public Diplomacy Division, NATO Programme for Security through Science, vol. 3 (IOS Press, 2006)

X. Jiang, N.Y. Chen, J.I. Hong, K. Wang, L. Takayama, J.A. Landay, Siren: context-aware computing for firefighting. Proc. Pervasive Comput. 18–23 (Apr 2004)

X. Jiang, J. Hong, L. Takayama, J. Landay, in *Ubiquitous computing for firefighters: field studies and prototypes of large displays for incident command*, Proceedings of ACM CHI Conference on Human Factors in Computing Systems, April 2004, pp. 670–686

G. Kantor, S. Singh, R. Peterson, D. Rus, A. Das, V. Kumar, G. Pereira, J. Spletzer, Distributed search and rescue with robot and sensor teams. Springer Tracts Adv. Robot. **24**, 529–538 (2006)

M.J. Karter, J.L. Molis, *Firefighter Injuries During 2008* (NFPA, Quincy, 2009)

M.J. Karter, *U.S. Fire Department Profile* (NFPA, Quincy, 2007)

J. Keggler, The Soldier as Nucleus. Armada Int. **34**(4), 42 (2010)

H.S. Kenyon, Soldiers' Tools Go Solar. Signal **61**(7), 67 (2007)

M. Klann, Tactical navigation support for firefighters: the LifeNet ad-hoc sensor-network and wearable system. Lect. Notes Comput. Sci. **5424**, 41–56 (2009)

S.V. Klimenko, V. Stanislav et al., Using virtual environment systems during the emergency prevention, preparedness, response and recovery phases. NATO Sci. Peace Secur. Ser. C: Environ. Secur. **6**, 475–490 (2008)

J.W. Knoll, A new trend in minified communications equipment. IRE Trans. Veh. Commun. **9**(1), 25–32 (1957)

A.R. Kohler, L.M. Hilty, C. Bakker, Prospective impacts of electronic textiles on recycling and disposal. J. Ind. Ecol. **15**(4), 496–511 (2011)

K.M. Kowalski, Electronics textiles wiring the fabrics of our lives. Odyssey **15**(6), 13 (2006)

L.D. Kozloski, *U.S. Space Gear: Outfitting the Astronaut* (Smithsonian Institution Press, 1994)

F. Krimgold, K. Critchlow, N. Uda-gama, *Emergency Service in Homeland Security*, NATO Security Through Science Series, 2006, pp. 193–229

E. Lee, S. Park, J. Lee, S. Oh, S. Kim, Novel service protocol for supporting remote and mobile users in wireless sensor networks with multiple static sinks. Wireless Netw. **17**(4), 861–875 (2007)

G. Leitner, D. Ahlström, M. Hitz, Usability of mobile computing in emergency response systems—lessons learned and future directions. Lect. Notes Comput. Sci. **4799**, 241–254 (2007)

A. Lymberis, Advanced wearable sensors and systems enabling personal applications. Lect. Notes Electric. Eng. **75**, 237–257 (2010)

R. Mandal, I. Singh, Optical design and salient features of an objective for a firefighting camera. Opt. Eng. **46** (Aug 2007)

J. Manyika, M. Chui, B. Brown, J. Bughin, R. Dobbs, C. Roxburgh, A.H. Byers, Big data: the next frontier for innovation, competition, and productivity, McKinsey Global Institute, May 2011, website: www.ciosummits.com/media/pdf/solution_spotlight/McKinseyGI_big-data-about.pdf, cited: 24 Feb 2012

D. Marculescu, R. Marculescu, Z. Stanley-Marbell, K. Park, J. Jung, L. Weber et al., Electronic textiles: a platform for pervasive computing. Pro. IEEE **91**(12), 1993–1994 (2003)

A. Mayhew-Smith, Soldiers get more punch. Electron. Wkly **2156**, 17 (2004)

S.L. Mazza, *An evaluation of self contained breathing apparatus voice communication systems*, EFO Paper, United States Fire Administration, May 2008

R.B. McNamee, *The use of personal alert safety devices to decrease levels of firefighter risk of death and injury*, EFO Paper, United States Fire Administration, Dec 1994

D. Merrill, Cranking up the heat on firefighters' radios. Mob. Radio Technol. **25**(7), 4 (2007)

Micro fuel processing system can power today's electronic soldier. Sci. Lett. **19** (2003)

MIL-PRF-28800F, Performance Specification, General Specification for Test Equipment with Electrical and Electronic Equipment, 24 June 1996, website: http://www.everyspec.com/MIL-PRF/MIL-PRF+(010000+-+29999)/MIL-PRF-28800F_18207/, cited: 30 January 2012

L.E. Miller, in *Indoor navigation for first responders: a feasibility study*, National Institute of Standards and Technology, Advanced Network Technologies Division, 10 Feb 2006, website: https://www.hsdl.org/?view&did=478117, cited: 6 Feb 2012

N. Moayeri, J. Mapar, S. Tompkins, K. Pahlavan, Emerging opportunities for localization and tracking. IEEE Wireless Commun. **18**(2), 8–9 (2011)

P. Möller, R. Loewens, I.P. Abramov, E.A. Albats, EVA suit 2000: a joint European/Russian space suit design. Acta Astronaut. **36**(1), 53–63 (1995)

L.N. Molino, Electronic control devices and EMS. Fire Eng. **161**(8), 30 (2008)

L.K. Moore, *Emergency Communications* (Nova Publishers, 2007)

M. Mordecai, Physiological status monitoring for firefighters. Firehouse **33**(9), 112 (2008)

R.R. Murphy, S. Tadokoro, D. Nardi, A. Jacoff, P. Fiorini, H. Choset, A.M. Erkmen, in *Search and Rescue Robotics,* Springer Handbook of Robotics, Part F, 2008, pp. 1151–1173

National Research Council, Panel on Human Factors in the Design of Tactical Display Systems for the Individual Soldier, in *Tactical Display for Soldiers: Human Factors Considerations* (National Academies Press, 1997)

NFPA 1500, *Standard on Fire Department Occupational Safety and Health Program,* 2007 edn. (National Fire Protection Association, Quincy, 2007)

NFPA 1801, *Standard on Thermal Imagers for the Fire Service,* 2010 edn. (National Fire Protection Association, Quincy, 2010)

NFPA 1981, *Standard on Open-Circuit Self-Contained Breathing Apparatus (SCBA) for Emergency Services,* 2007 edn. (National Fire Protection Association, Quincy, 2007)

NFPA 1982, *Standard on Personal Alert Safety Systems (PASS)*, 2007 edn. (National Fire Protection Association, Quincy, 2007)

A.S. Park, S.S. Kim, in *A novel agent-based user-network communication model in wireless sensor networks*, Networking '07, Proceedings of the 6th International IFIP-TC6 Conference on Ad-hoc and Sensor Networks, Wireless Networks, Next Generation Internet, 2007

J. Pallauf, P. Gomes, S. Bras, J.P. Cunha, M. Coimbra, Associating ECG Features with Firefighter's Activities, *Annual Conference Proceedings of the Conference of the IEEE Engineering in Medicine and Biology Society*, 6009-12, Aug 2011

PASS Signals Can Fail at High Temps. Fire Chief, IAFF **49**(12): 10, (2005)

B.A. Plog, P.J. Quinlan, *Fundamentals of Industrial Hygiene*, 5th edn. (National Safety Council, 2002)

A.D. Putorti Jr, F.K. Amon, K.M. Butler, C.A. Remley, W.F. Young, C. Spoons, *Structural and electromagnetic scenarios for firefighter locator tracking systems*, NIST TN 1713, National Institute of Standards and Technology, Gaithersburg, MD, 2011

Proposed Standard on Electronic Safety Equipment for Emergency Services, Pre-ROP Draft, National Fire Protection Association, Quincy, MA, 17 Nov 2006

L. Ramirez, T. Dyrks, J. Gerwinski, M. Betz, M. Scholz, V. Wulf, Landmarke: an ad hoc deployable ubicomp infrastructure to support indoor navigation. Pers. Ubiquit. Comput. **10** (2007)

J. Rantakokko, J. Rydell, P. Stromback, P. Handel, J. Calmer, D. Tornqvist, F. Gufstafsson, M. Jobs, M. Gruden, Accurate and reliable soldier and first responder indoor positioning: multisensor systems and cooperative localization. IEEE Wireless Commun. **18**(2), 10–18 (2011)

M. Richardson, R. Scholer, Thermal imaging training: covering the basics. Firehouse **26**(4), 86–88 (2001)

M.R. Roberts, NIST tests firefighter tracking devices for radio-frequency interference. Urgent Commun. (2011)

M.R. Roberts, Wireless sensor system guides urban firefighters. Urgent Commun. 1 Dec 2006, website: http://urgentcomm.com/mag/radio_wireless_sensor_system/, cited: 10 Feb 2012

M. Ruta, E. DiSciascio, F. Scioscia, RFID-Enhanced Ubiquitous Knowledge Bases: Framework and Approach, *Unique Radio Innovation for the 21st Century, Part 3*, 2010, pp. 229–255

C. Sapateiro, P. Antunes, G. Zurita, R. Vogt, N. Baloian, Evaluating a mobile emergency response system. Lect. Notes Comput. Sci. **5411**, 121–134 (2008)

K. Shreeves, Winner: Poseidon discovery—tools & toys. IEEE Spectr. **46**(1), 28–29 (2009)

A. Skoog, The EVA space suit development in Europe. Acta Astronaut. **32**(1), 25–38 (1994)

E.J. Spahn, *Fire Service Radio Communications* (Fire Engineering, NY, 1989)

D. Steingart, J. Wilson, A. Redfern, P. Wright, R. Romero, L. Lim, *Augmented Cognition for Fire Emergency Response: An Iterative User Study*, University of California Berkeley, paper presented at Proceedings of Augmented Cognition, HCI International Conference, 2005

Y. Sun, K.Y. Ong, *Detection Technologies for Chemical Warfare Agents and Toxic Vapors* (CRC Press, 2005)

L.M. Sunderman, Improving Firefighter Accountability Systems with the Use of Electronic Devices, Executive Fire Officer Paper, U.S. Fire Administration National Fire Academy, Emmitsburg, MD, June 2006

Tanagram, About Us, website: http://tanagram.com/2012/01/24/defining-augmented-reality-and-why-it-will-be-a-performance-enhancing-technology/, cited: 24 Jan 2012

B.W. Teele, Fire and emergency services protective clothing and protective equipment, *Fire Protection Handbook*, 20th edn., Section 12, Chapter 9 (National Fire Protection Association, Quincy, 2008) pp. 12-143–12-160

Y.C. Tsao, L.C. Chen, S.C. Chan, The research of using image-transformation to the conceptual design of wearable product with flexible display. Lect. Notes Comput. Sci. **4551**, 1220–1229 (2007)

D. Tuite, Radio Interoperability: it's harder than it looks. Electron. Des. **56**(8), 30 (2008)

UL 913, *Intrinsically Safe Apparatus and Associated Apparatus for Use in Class I, II, and III, Division 1, Hazardous (Classified) Locations*, Underwriters Laboratories, Northbrook, IL, 31 July 2012

USDHS, *Where There's Smoke, There's a Signal*, U.S. Department of Homeland Security, Science and Technology Directorate, July 2011, website: www.dhs.gov/files/programs/st-snapshots-self-powered-waterpoof-heat-resistant-router.shtm, cited: 31 July 2012

USFEMA, *In Development Technologies*, Responder Knowledge Base, website: www.rkb.us/technologies.cfm, cited: 30/July 2012

Voltage Regulator Tutorial & USB Gadget Charger Circuit, Afrotechmods, website: http://www.youtube.com/watch?v=GSzVs7_aW-Y, cited: 20 Feb 2012

J.J.H. Wang, J.K. Tillery, K.E. Bohannan, G.T. Thompson, Helmet mounted smart array antenna. IEEE Int. Symp. Antennas Propag. **1**, 410–413 (1997)

L. Wei, Z. Zhang, Q. Wang, A. Feng, The research and implementation based on electronic gases simulation system of wireless sensor network. Adv. Mater. Res. **433**(440), 2519–2522 (Jan 2012)

C.L. Werner, Fire service technology: looking ahead. Firehouse **25**(7), 118 (2000)

C. Whitby, Firefighter ergonomics enhance equipment performance. Fire Eng. **163**(6), 101 (2010)

R.C. Wilde, J.W. McBarron, S.A. Manatt, H.J. McMann, R.K. Fullerton, One hundred US EVAs: a perspective on spacewalks. Acta Astronaut. **51**(1), 579–590 (2002)

J. Wilson, V. Bhargava, A. Redfern, P. Wright, *A Wireless Sensor Network and Incident Command Interface for Urban Firefighting*. Fourth Annual International Conference on Mobile and Ubiquitous Systems: Networking & Services, Aug 2007

J. Wilson, D. Steingart, R. Romero, J. Reynolds, E. Mellers, A. Redfern, L. Lim, W. Watts, C. Patton, J. Baker, P. Wright, Design of monocular head-mounted displays for increased indoor firefighting safety and efficiency. Proc. SPIE, **5800** (2005)

Wireless System for Monitoring Weather Conditions in Wildland Fire Environments. Paper presented at Proceedings of the 4th International Conference on Mobile Systems, Applications and Services, 2006

Zephyr, website: www.zephyr-technology.com/, cited; 1 March 2012